Owning the Green Grid

American and Comparative Environmental Policy

David M. Konisky, Megan Mullin, Kate O'Neill, series editors

A list of titles in this series appears at the back of the book.

Owning the Green Grid

The Political Economy of Renewable Energy Policy Design

Joshua A. Basseches

The MIT Press
Cambridge, Massachusetts
London, England

The MIT Press
Massachusetts Institute of Technology
77 Massachusetts Avenue
Cambridge, MA 02139
mitpress.mit.edu

The MIT Press would like to thank the anonymous peer reviewers who provided comments on drafts of this book. The generous work of academic experts is essential for establishing the authority and quality of our publications. We acknowledge with gratitude the contributions of these otherwise uncredited readers.

This book was set in Stone Serif and Stone Sans by Westchester Publishing Services. Printed and bound in the United States of America.

Library of Congress Cataloging-in-Publication Data is available.

ISBN: 978-0-262-05358-7

EU Authorised Representative: Easy Access System Europe, Mustamäe tee 50, 10621 Tallinn, Estonia | Email: gpsr.requests@easproject.com

To my parents

Contents

Acknowledgments

This book has been nearly eight years in the making. It would not have been possible without the generosity, time, and assistance of archivists, librarians, and, of course, 219 state-level policymakers and energy policy professionals, who collectively spent hundreds of hours of their precious time teaching me a great deal of what I know about electricity politics and policy.

The book began as a dissertation project at Northwestern University, where I received first-rate advising and mentorship from Monica Prasad and Tony Chen. Monica always encouraged me to ask bigger questions (and study more states!), to use research to solve social problems rather than just to describe them, and to pursue climate policy as a topic of study. Tony helped sharpen my thinking about corporate political power, and read my work line by line, pushing me to make it stronger and to make my arguments airtight. They were both selfless in their dedication to advising me and together helped me develop into the scholar I am today.

Tony introduced me to Barry Rabe in graduate school, who ended up becoming an indispensable mentor and asset to my early career as a faculty member. After providing helpful input as a member of my dissertation committee, he brought me on as a postdoc at the Ford School of Public Policy at the University of Michigan, where I learned all I know about Canadian and Mexican climate politics in conjunction with our joint project: the North American Colloquium on Climate Policy. Through this project, I was introduced to other leaders in the field. Meanwhile, Barry supported me in my pursuit of funding to complete the research for this book, and on the job market that brought me to Tulane.

Brayden King and Heather Schoenfeld also served on my dissertation committee, and I am grateful to Brayden for our conversations about the political influence of social movements and to Heather for hiring me as an RA to work on her project with Michael Campbell on state-level incarceration policies as well as for her collaboration on methodological approaches to studying state-level politics and policymaking. David Cunningham will always have the distinction of launching me on my academic career as my undergraduate thesis advisor at Brandeis University. As I now advise undergraduate thesis students of my own, I try to emulate David's approach.

I was fortunate to meet Timmons Roberts (virtually) in 2020. Timmons is the executive director of the Climate Social Science Network (CSSN), which supported the research in this book. Through Timmons, I became involved in CSSN, where I cochair the State Politics Working Group along with Rebecca Bromley-Trujillo, who also deserves my thanks for being a terrific partner in these efforts and a wonderful colleague and coauthor. In addition to CSSN, I thank the various other funders of the different stages of this project, including the Renewable Energy Policy Initiative at the University of Michigan's Ford School, a Lavin-Bernick Grant and SLA Faculty Research Award at Tulane University, a Graduate Research Grant and Comparative Historical Social Science Grant from Northwestern University, and a Centennial Center Research Grant from the American Political Science Association.

I am grateful for feedback that I received at various stages of this project from audiences at several academic conferences and workshops. These include annual meetings of the American Sociological Association (ASA), the Southern Political Science Association (SPSA), the Association for Public Policy Analysis and Management (APPAM), and the Social Science History Association (SSHA), as well as the Conference on Policy Process Research (COPPR), the Political Economy of Climate and the Environment (PECE) Conference, the Consortium on the American Political Economy (CAPE) Junior Working Group, the Conference on Politics of Clean Energy Abundance Under Federalism at Princeton, the Climate Pipeline Project workshop also at Princeton, and likely others as well. Katie Lamb at Catalyst Cooperative helped with the data on utility parent companies and subsidiaries.

I received valuable input from several colleagues, including (but perhaps not limited to—apologies to anyone I may be forgetting!) Rebecca

Bromley-Trujillo, Sanya Carley, David Carter, Alex Garlick, Amy Jeon, Timmons Roberts, Dustin Tingley, Sam Trachtman, and members of the CSSN State Politics Working Group. Michael Kraft and Sheldon Kamieniecki went above and beyond, reading and providing extremely helpful and detailed comments on the entire manuscript. I cannot thank them enough. My thanks as well to the ACEP Series Editors, David Konisky—who I enjoyed crossing paths with at multiple conferences while working on the manuscript—Megan Mullin, and Kate O'Neill. I also thank the anonymous peer reviewers, and Beth Clevenger, who has been a delight to work with at the MIT Press. Any remaining errors are mine alone.

My "junior colleagues" group at Tulane has been extremely supportive and helpful throughout; this group includes Brandon Davis, Andrew Leber, Rachel Schoner (my "hurricane buddy"), Allison Grossman, Emily Rains, Stan Oklobdzija, and Maayan Mor. Additionally, a few senior scholars in my department have gone out of their way to provide mentorship including Mirya Holman (now at University of Houston), Nancy Maveety, Moises Arce, and Thomas Oatley. Outside my department, Jesse Keenan has been very supportive. My department chair, Martin Dimitrov, has been extremely supportive, as has the program director of environmental studies, Laura McKinney, in making my joint appointment at Tulane work out smoothly. I am grateful for the support of Tulane University's School of Liberal Arts.

This book would not have come to fruition without the help of six outstanding undergraduate research assistants, at Northwestern University, University of Michigan, and Tulane. At Northwestern, Sarah Kulaga and Kaitlin Rubinstein were instrumental in getting things off the ground. At University of Michigan and Tulane, Zona Martin, Drew Cox, Libby Antonneau, and John Lockwood were outstanding. These four were knee-deep in legislative text, PUC filings, and many, many spreadsheets. I was impressed with what fast learners they were and feel very lucky to have had all six of them on my team.

Many friends kept me grounded in "real life" as I toiled for many years over the ups and downs of this book project. There are too many to name all of them, but they include Drake, Bob, Erik, Shawn, Daniel, Ben, Leah, Ariel, Mark, Jacob, Herbie, Maddie, Hannah, Abby, and Yoni. Jason Brown has been my "best friend" and hiking buddy since 2020, and I could not have asked for a better partner as we both navigate the pressures of academia together, as well as the best moments in life. Jason has pushed me and helped me grow, both personally and professionally.

My parents, Ruth Aronson and Michael Basseches, to whom this book is dedicated, deserve the greatest thanks of all. They have been there for me every step of the way, from birth to book. My grandparents, although no longer with us, have had a profound impact on who I am today. My brother, Benya, his wife, Eileen, as well as Alan, Angela, Uncle David, and Lynne have my sincere appreciation as well.

1 The Political Economy of the Electricity Grid

> We weren't afraid of the technology. We were ready to embrace an RPS [renewable portfolio standard].
> —author's interview, 2/26/18

It may surprise readers to learn that these words were spoken by a corporate lobbyist for an investor-owned utility (IOU) company. The setting was the state of Oregon, and the lobbyist was referring to his company's support for Senate Bill 838, which was signed into law by that state's Democratic governor, Ted Kulongoski, in 2007. A different Oregon IOU lobbyist, employed by a different company, explained, "Renewables are good for utilities. . . . They're capital intensive. Fuel is free. You don't earn on fuel costs anyway. . . . For the traditional business model, which is invest in capital and earn a regular return on it, that's a good thing" (author's interview, 2/19/18).

This insight about the IOU business model, and how unusual it is, turns out to have major implications for the interest group politics of state-level climate and renewable energy policymaking. This book, among other things, will be about what those implications are and how they depend on the particular state context. Yet before we delve into that, we need to understand something else, related to this anecdote from Oregon: Why would an IOU not only *not oppose* a state-level RPS bill but actively *support* it?

Here, additional quotes from these same two lobbyists are instructive. The first is in response to a question asking the lobbyist to elaborate on why they preferred the renewable portfolio standard (RPS) be adopted through legislation rather than a ballot initiative. A ballot initiative campaign had been threatened by Oregon environmental advocates just prior to the

2007 legislative session, in the aftermath of a successful RPS ballot initiative adopted in 2006 in Washington State, Oregon's neighbor to the north. The second quote relates to the process of negotiating the Oregon RPS bill's substantive content:

> Legislators tend to be very reluctant to change things that were adopted by [ballot] initiative. I think that almost every session since the RPS was adopted [legislatively], it's been amended [by the legislature]. (author's interview, 2/19/18).

> The politics were going there, and so in that fall we were told this is coming, please work with us. And so we pulled on some of our expertise from Iowa and within our holding company to craft really five basic principles, or five basic needs, must-haves, in that bill. (author's interview, 2/26/18)

In other words, Oregon IOUs realized the inevitability of this policy change—Oregon policymakers had resolved to adopt an RPS one way or another—and yet, these IOU lobbyists saw an opportunity to craft the legislative language to their benefit. They were invited to the table to negotiate the bill to their liking, and they far preferred that approach to it being "shoved down their throats" by the state's progressive voters (who also happened to be these companies' captive electricity customers). Although, as we will see later in this book, the structure of Oregon's electric utility sector was not especially conducive to IOUs' embrace of an RPS, and one Oregon IOU in particular was initially quite resistant, enough other political forces (including the threat of a ballot initiative) brought IOUs to the negotiating table. The second lobbyist's comment about Iowa and his employer's "holding company" is also relevant and reflects a second way—besides the business model—that the electric utility industry is highly unusual. It is often characterized by what political sociologists Prechel and Morris (2010, 337) call the "multi-layer subsidiary form," and we will get into how that matters more as well.

For the time being, though, this anecdote from Oregon raises some important questions for social scientists interested in interest group politics, business power, state politics, and the quality of democratic representation (e.g., Anzia 2019; Chen 2019; Grumbach 2022; Hertel-Fernandez 2019; Hacker, Hertel-Fernandez, Pierson, and Thelan 2021; Kamieniecki 2006, 1): How typical or unusual is this calculation on the part of a business actor, that it is better to negotiate the policy substance and ultimately support a piece of legislation than to oppose it outright? Relatedly, 2) How typical or unusual is it that a business actor recognizes the inevitability of legislative adoption at the outset of the legislative process? 3) How typical or unusual is it that a business actor assumes the ability to successfully condition support on as many as "five basic needs, must-haves?" Relatedly, 4) If a particular business actor *is* granted this ability, at whose expense does it come? 5) Are certain types of business actors granted this ability more routinely than others, and, if so, what are the characteristics that define these "lucky ones"?

Undoubtedly, the answers to all these questions will vary based on the legislative context and the bills in question. Instead of providing a universally correct or generalizable answer to any of these questions, this book will 1) illuminate why IOUs are unique corporate political actors in the area of state-level renewable electricity policymaking; 2) show how IOUs' policy preferences differ from those of competing political actors in this same policy space; 3) demonstrate how, and argue as to why, IOUs tend to get their way more often than any of the other competing actors; 4) show how the particular state context matters in terms of IOUs' policy preferences; and 5) explain what it all means for prospects for US climate and renewable energy policy going forward.

The remainder of this introductory chapter provides empirical background information on the history, governance, and structures of the United States electricity sector. It provides an overview of the types of political actors routinely engaged in electricity policymaking, and it highlights why IOUs are unique among these actors in terms of how their business models and corporate structures affect their policy preferences regarding renewable energy mandates. It introduces state-level utility sector structure as an important explanatory variable for understanding IOUs' preferences regarding RPS policies, and it sets us up to explore in the subsequent chapters how all of this informs RPS policy design.

US Electricity Sector History, Governance, and Structure

This book argues that IOUs are a unique type of political actor when it comes to shaping the design of state-level RPS policies. To understand why, we must understand IOUs' history, how they are structured in the contemporary period, how electricity (the service IOUs provide) is governed at multiple levels in the US system, and finally, how the structure of the electricity sector itself varies by state.

Electric Utility Industry History and Corporate Structure

The electric utility industry has evolved from its origins around the turn of the twentieth century to become one of the most complex and unusual private industries in existence today. It is characterized by inelastic consumer and business demand for its product (electricity, which has, of course, become an essential piece of modern American life), by its corporate structure involving multilayered subsidiaries (recall the lobbyist's reference to the "holding company"), and by its privileged status in both the public and private sectors, with IOUs being government-regulated corporate monopolies. "Regulation" might not *sound* like a "privilege," but as we will see in this book and as has been shown elsewhere (e.g., Kovvali and Macey 2022; Pomerantz 2023), in IOUs' case, it certainly is, especially since that regulation has been accompanied by the guarantee of monopoly (Basseches 2024; Hempling 2020; Hirsh 1999; Isser 2015; Vaheesan 2024).

The IOU industry mastered the game of politics from its earliest days. Thomas Edison, the inventor of the modern electricity system and founder of what eventually became the New York utility company, Consolidated Edison (ConEd), hired Samuel Insull, first as his personal secretary and then as his chief financial strategist. Insull is described as "a master of public relations" (Hughes 1983, 201). In 1892, Insull left his job at General Electric, a company worth $50 million at the time (a lot of money in those days), to become president of Chicago Edison, Chicago's local electric utility and the predecessor of Commonwealth Edison (ComEd), which was worth just $885,000 when Insull arrived. Over the next two decades, Insull transformed the company, creating "a monopolistic, mass-producing, technologically efficient, and economically operated company for all of Chicago" (Hughes 1983, 204), very much working with local and state-level politicians to do so (Hughes 1983).

Consequently, Insull's business success became a model for his industry contemporaries, and Insull became "a spokesman for the utilities industry" (Hughes 1983, 204). Insull's biographer described him as "the most powerful political operator on the American business scene" (Hughes 1983, 205) for a period of about thirty-five years, before he became mired in scandal, eventually charged with mail fraud and blamed by his critics for instigating the Great Depression (Hughes 1983; Rillinger 2019). Today, Insull's former company, ComEd, is a subsidiary of Exelon Corporation, which owns ComEd and five other IOU subsidiaries operating in five states plus

Washington, DC. The parent company employs 18,000 and owns assets valued at more than $95 billion, placing it in the top 100 US corporations (Exelon Corporation 2023; Fortune Magazine 2023).

One of Insull and his cohorts' most ingenious political maneuvers was establishing their business as a "natural monopoly" in the eyes of the government and establishing state governments, rather than the federal government, as their primary regulator. As IOU businesses grew, they developed both adversaries and skeptics. Their biggest adversaries were publicly owned utilities (POUs), which, at the time, prior to the rural electrification movement, were mostly municipally owned utilities in significant population centers (Vaheesan 2024). Their biggest skeptics were their ratepayers, who were concerned that corporate greed was inflating their rates. Government was bound to get involved to help settle this tension. Insull and his cohorts knew this, and so they proactively proposed to lawmakers a deal which became known as the "regulatory compact," or as energy historian Richard F. Hirsh (1999) calls it, "the utility consensus."

The idea was simple and steeped in a preexisting legal paradigm that governed gas companies and railroads (Vaheesan 2024): Given the enormously capital-intensive physical infrastructure that was required to produce and distribute electricity, competition among corporate actors in a given geographic territory, or even between IOUs and POUs, was inefficient and impractical. Therefore, IOUs should be allowed to operate as monopolies in defined geographic service territories. In exchange for this privilege, they would subject themselves to regulation by state government. State governments would establish public utility commissions (PUCs), which would review and approve company-proposed electric utility rates, since, unlike other commodities, electricity rates would not be determined by free market forces but rather by capital expenditures. To help these IOUs attract capital, PUCs would not only allow them to pass all of their "used and useful" capital costs on to their captive customers, but they would also allow them to collect from these customers a guaranteed "fair rate of return" on all of their investments. This process by which rates would be determined became known as "cost-of-service" regulation (Gormley 1983; Hirsh 1999; Vaheesan 2024).

The first PUCs were established in New York and Wisconsin in 1907. By 1914, forty-three other state PUCs had been created, and soon after the remaining states established them as well, eventually appearing in all fifty states.[1] These PUCs were originally created for the primary purpose of balancing the interest of IOUs in attracting capital with the interest of ratepayers, who would otherwise be at the total mercy of these profit-seeking monopoly corporations. However, their establishment had broader political consequences. For one thing, as Vaheesan writes, "The stuff of democratic debate and decision making became, in part, technocratic matters removed from popular politics" (2024, 34). For another, the more formalized standing granted to IOUs dealt a blow to the municipalization movement, and the movement for POUs more generally. Both these consequences of the establishment of the regulatory compact advantaged IOUs and their investors politically and were likely foreseeable by industry managers like Insull (Vaheesan 2024).

Over the years since PUCs were first created, state legislatures have delegated an increasing number of additional regulatory and policymaking powers and responsibilities to them—everything from approving utility mergers and acquisitions to investigating quality-of-service issues to, of course, implementing RPS policies. Moreover, electricity became just *one* of the industries PUCs were responsible for regulating: Depending on the state, other industries falling under PUC jurisdiction included oil and gas (for heating and transportation), telecommunications, railroads, taxis and limousines, and water (Gormley 1983; Heern 2023; Parinandi 2023).

Under the regulatory compact, the electric utility industry grew increasingly large, lucrative for its investors, and concentrated in its wealth. Holding companies, which unlike operating utilities were *not* rate-regulated and had footprints that transcended state borders and even regional geographies, were ostensibly stood up to help finance the high up-front cost of capital needed to build out a nationwide electricity system. They ended up providing a vehicle to establish "industry supremacy" (Vaheesan 2024, 34), including "hidden value transfers" between operating companies that evaded regulatory scrutiny (Kovvali and Macey 2022). In 1930, three quarters of the electricity sold in the United States was accounted for by just ten holding companies. The Public Utility Holding Company Act (PUHCA) of 1935 was an effort to rein in these corporate pyramids, and succeeded to a degree, but fell short of eliminating the multilayered subsidiary structure, with the IOU industry being extremely active in influencing the legislative language (Vaheesan 2024).

By 2021, there were 167 electric IOU companies operating in the fifty states, but seventy-five of these (roughly 45 percent) were still owned by just twenty-two holding companies—all Fortune 500 companies with combined assets valued in excess of $2.5 trillion. The largest of these, in terms of asset value, were Berkshire Hathaway (owning five regulated IOUs operating in ten states), Duke Energy (owning six regulated IOUs operating in six states), and Next Era Energy (owning two regulated IOUs both operating in Florida). Examples of other major players include Southern Company, Dominion Energy, American Electric Power, and the previously mentioned Exelon. It is notable that many of these same energy holding companies own nonregulated subsidiaries as well, and oftentimes the regulated and nonregulated subsidiaries do business with one another. Table 1.1 lists the largest IOUs operating in the seven states covered in greatest detail in this book, along with their holding companies, the number of IOUs owned by those holding companies, and examples of nonregulated businesses owned by the same holding company (US Energy Information Administration 2021; Fortune Magazine 2023).

Table 1.1
Corporate relatives of largest IOUs in case study states, as of 2021

State	Largest Regulated Electric IOUs	Parent Companies (# of regulated IOUs owned across all states)	Other Subsidiary Example
CA	Pacific Gas and Electric	PG&E Corp. (1)	*N/A*
	San Diego Gas and Electric	Sempra Energy (2)	Sempra Infrastructure
	Southern California Edison	Edison International (1)	*N/A*
MA	Eversource	Eversource Energy (3)	Hopkinton LNG Corp.
	National Grid	National Grid (1)	National Grid Partners
OK	Oklahoma Gas and Electric	OGE Energy Corp. (1)	*N/A*
	Public Service Co. of Oklahoma	American Electric Power (10)	AEP OnSite Partners
OR	Pacific Power/PacifiCorp	Berkshire Hathaway (5)	BHE GT&S
	Portland General Electric	PGE Corp. (1)	*N/A*

(*continued*)

Table 1.1
(continued)

State	Largest Regulated Electric IOUs	Parent Companies (# of regulated IOUs owned across all states)	Other Subsidiary Example
PA	Penn Power*	Allegheny Energy*	*N/A*
	Duquesne Power and Light	DQE Holdings (1)	*N/A*
	Metropolitan Edison (Met-Ed)	First Energy (9)	FirstEnergy Service Co.
	Pennsylvania Electric (Penelec)	First Energy (9)	FirstEnergy Service Co.
	Pennsylvania Power and Light	PPL Corp. (3)	PPL Capital Funding, Inc.
	Philadelphia Electric Co.	Exelon (7)	*N/A***
	UGI Utilities, Inc.	UGI Corp. (1)	UGI Energy Services
	West Penn Power	First Energy (9)	FirstEnergy Service Co.
SC	Duke Energy Carolinas	Duke Energy Corp. (6)	Duke Energy One
	Dominion South Carolina	Dominion Energy (2)	Domin. Energy Solutions
TX	Houston Electric	CenterPoint Energy (4)	Vectren Affiliated Utilities
	El Paso Electric	El Paso Electric (1)	*N/A*
	Entergy Texas	Entergy (4)	System Energy Resources
	Oncor Electric	Oncor Electric (1)	Oncor Receiv-ables, LLC
	Southwestern Public Service Co.	Xcel Energy (4)	WestGas Inter-State Inc.
	Texas-New Mexico Power	PNM Resources (2)	TNP Enterprises, Inc.

*Acquired by First Energy in 2010.

**Exelon recently split into two separate, publicly traded companies, one for its regulated utility businesses and one for its nonregulated energy production and customer-facing energy services businesses.

Note: Some of the information in this table was verified with the help of Catalyst Cooperative, which maintains a database that relies on data science and machine learning techniques to link unstructured data from the US Securities and Exchange Commission (SEC) and the US Energy Information Administration (EIA).

The Electricity System and Its Governance

IOUs and their holding companies operate across a complex electricity system that includes three major levels: 1) electric generation, 2) electric transmission, and 3) electric distribution. "Generation" involves the production of electricity from a variety of extracted and refined fuels in the case of fossil-fueled electricity, or alternatively, from renewable sources in the case of wind, solar, geothermal, hydropower, and other forms of "renewable electricity." "Transmission" involves the transportation of those electrons once generated, regardless of their source, to local substations. This is achieved through high-voltage transmission lines that may or may not cross state political borders.[2] Ultimately, these transmission lines serve as the skeletons of three internally interconnected electricity "grids" that together electrify the continental United States (and large portions of Canada and Mexico): the Eastern Interconnection, the Western Interconnection and the Texas Interconnection[3] (Cohn 2019). And finally, "distribution" involves getting electricity from the substations into the homes and businesses of electricity customers[4] (US Department of Energy 2017).[5]

In all fifty states, state-level PUCs regulate the second two levels (transmission and distribution) but whether they also regulate the first level, generation, depends on the governance and regulatory structure of the electric sector, which varies by state and region in ways we will soon discuss. The regulation of distribution is exclusively the province of state governments and their PUCs. The regulatory responsibilities for transmission, however, are shared between state PUCs and the federal government, specifically the Federal Energy Regulatory Commission (FERC). This is because many transmission lines cross state borders (Borenstein and Bushnell 2015).

In between state and federal governments, in some but not all regions of the country, are quasi-governmental bodies known as independent system operators (ISOs) or regional transmission organizations (RTOs). These were authorized by FERC Orders 888/889 in 1996 and FERC Order 2000 in 1999, respectively, around the same time certain state legislatures began restructuring their electric sectors (more on that soon). In plainer language, ISOs/RTOs function as the "grid operators." They were created as part of a broader effort to move toward a more geographically integrated, and more competitive, wholesale electricity market (Borenstein and Bushnell 2015; Isser 2015).

ISOs and RTOs perform a similar function, and while there are some technical differences, they can be thought of as one and the same for the purposes of this book. They are both FERC-sanctioned, regionally based, nonprofit organizations that were created to coordinate and in many ways "referee" wholesale electricity market transactions in real time between an increasing number of market participants, each with their own financial self-interest, ideally ensuring nondiscriminatory pricing, fair access to, and optimal functioning of regional grids (Rilinger 2024). In essence, while they have no direct accountability to ordinary voters or electricity consumers,[6] which can be viewed as a fatal flaw from an "energy democracy" perspective (Burke and Stephens 2017), they operate with the consent of all electricity market participants, recognizing the need for a technically competent and financially independent arbiter of wholesale market transactions that can ensure that any disputes that may arise between market participants are resolved without resulting in power system outages (Costello and Burns 2000; Rose et al. 1997). Unfortunately, in practice, the largest corporate actors in their geographic footprints wield disproportionate influence over their governance (Peskoe 2023).

However, as mentioned previously, ISOs/RTOs do not appear uniformly throughout the entire country; they appear in large parts of the country where electricity market stakeholders elected to move toward wholesale competition and electricity market liberalization, often with the buy-in of state legislatures and PUCs, but this was not everywhere. In particular, the Southeast region and large parts of the West have been the holdouts (Harrison and Welton 2021). In these places, IOUs remain fully "vertically integrated" (exerting monopoly control over all three levels of the electricity system), thus precluding the need for any sort of independent arbiter of wholesale market transactions like an ISO/RTO. In these places, the IOUs *are* the grid operators and make their own rules, with some limited oversight from PUCs and FERC (Tsoukalis et al. 2023).

The Electric Reliability Council of Texas (ERCOT) is an ISO that governs the electricity market in most, though not all, of Texas. It is distinct from other ISOs/RTOs in that its geographic footprint is entirely contained within the political borders of a single state (Texas) *and* it has extremely minimal interconnection with surrounding grids (recall the three main grids—Eastern Interconnection, Western Interconnection, and, separately, Texas Interconnection). This means that ERCOT market participants escape oversight from FERC, and it also means that the generation resources available to the ERCOT control area of the electric grid are generally limited to within Texas, which is one (but not the only) reason that Texas experienced major reliability problems in recent years when electricity demand surged due to extreme weather events (Postelwait 2022). Figure 1.1 shows a map of the ISOs/RTOs in the continental United States, and table 1.2 lists those ISOs/RTOs that interact with the seven case study states in this book (US Environmental Protection Agency 2023).

Restructuring and Electricity Market Liberalization

The FERC orders that authorized the formation of ISOs/RTOs were part of a broader movement—involving both the federal government and the states—toward electricity market liberalization, also known as "restructuring" or "deregulation." In this book, I will use the term "restructuring," because it is more succinct than "electricity market liberalization" and more accurate than "deregulation." Although the term "deregulation" is often used, it is a misnomer. Its users are referring to the end of the era of vertically integrated IOUs (which, recall, remain in place in the Southeast

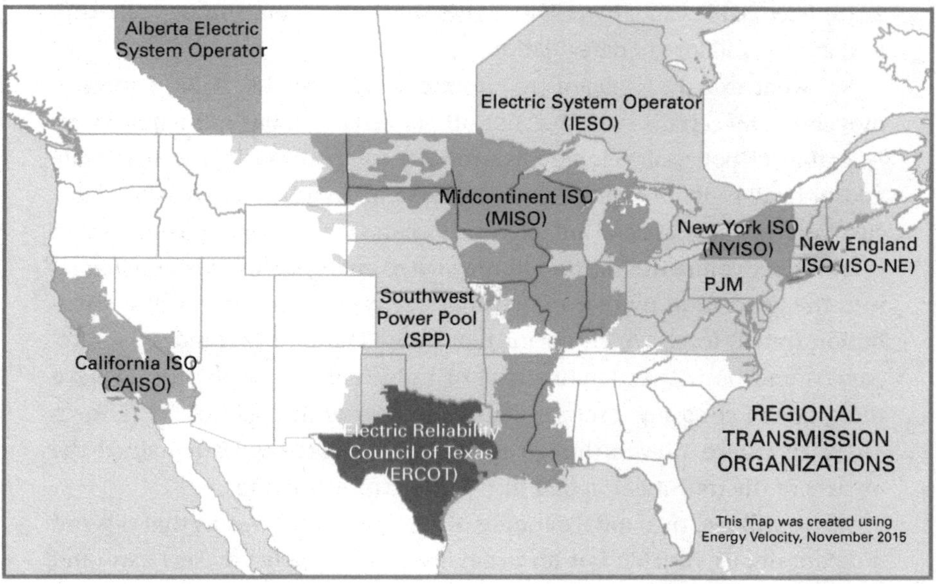

Figure 1.1
Map of geographic footprint of ISOs/RTOs in continental United States. *Source*: https://www.epa.gov/green-power-markets/power-market-structure

Table 1.2
ISOs/RTOs in case study states

State	ISOs/RTOs
CA	California ISO (CAISO)
MA	New England ISO (ISO-NE)
OK	Southwest Power Pool (SPP)
OR	*None; vertically integrated state*
PA	PJM Interconnection (PJM)
SC	*None; vertically integrated state*
TX	ERCOT; SPP; Midcontinent ISO (MISO)

and large portions of the West) and of PUC-regulated generation (which remains in place in these same states plus some others, like Oklahoma, which are part of an ISO/RTO); users of the term "deregulation" often (unintentionally) forget that the transmission and distribution levels of the electricity system remain regulated by the PUCs in all fifty states, as well as by FERC in the case of interstate transmission (Borenstein and Bushnell 2015; Isser 2015). Cudahy (1998) agrees this term is a misnomer, referring to it as "the folklore of deregulation."

So, what exactly is meant by "restructuring"? In this book, I mean a movement by certain states but not others—enabled, but not required, by federal-level policy developments—toward competition in the wholesale and/or retail functions of the electricity system. In other words, when I discuss "restructured" states, I am referring to states whose policymakers elected to introduce competition (from nonregulated energy sector business actors) into the generation (wholesale) level of the system and/or into the distribution (retail) level, which meant that individual homes (residential ratepayers) and businesses (commercial or industrial ratepayers) would have the *option* of choosing a nonregulated provider of their electricity (known as "retail choice"), even though monopoly regulated IOUs remained the owners of the transmission and distribution infrastructure.

As we will see, this understanding of "restructuring" means that it is not a dichotomous variable but an ordinal one; state policymakers can (and have) pursued different *degrees* of restructuring. Every step, or attempted step, further down the path toward "complete restructuring" has been fraught with political controversies among powerful energy sector business actors (Harrison and Welton 2021). In each state, policymakers have chosen the flavor of restructuring (including none at all) that the conditions of the state's political economy have allowed. In the case of the states that have gone the farthest, like Texas, this has meant "complete restructuring" at both the wholesale and retail levels, whereas in the states that have gone the least far, like South Carolina, this has meant the continuation of the status quo: vertically integrated IOU monopolies controlling all levels of the electricity system (Tsoukalis et al. 2023).

As we will see, there are many different policy design features of state-level electricity restructuring policies; however, two important ones are especially worth outlining here because they are critical to the interest group politics of RPS policy design that will be the focus of the chapters that follow: retail competition and generation divestiture. "Retail competition" or "retail choice" means that individual electricity customers (homes and businesses) have the *option* to choose a firm other than the incumbent IOU whose service territory they are geographically located in to source their electricity for them. Figure 1.2 shows the states that offer retail choice.

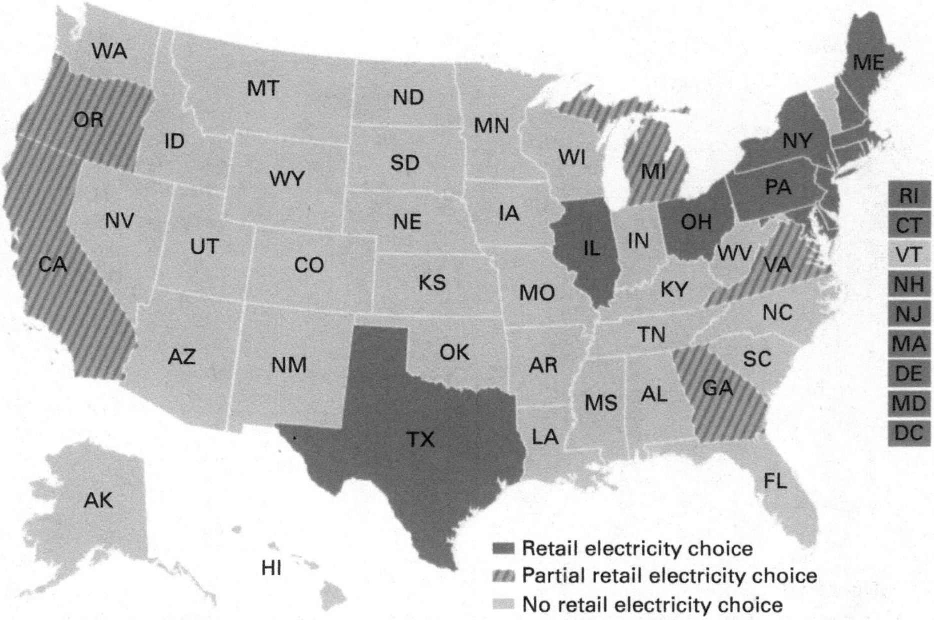

Figure 1.2
States with retail electricity choice. *Source*: https://www.epa.gov/green-power-markets/power-market-structure

As the figure shows, a few states have "partial retail choice," a nuance that can be put aside for the time being, but one that further illustrates the non-dichotomous nature of state-level restructuring policies.

The other important policy feature is whether restructured states *required* their IOUs to *divest their assets* from the generation level of the electricity system as part of their restructuring policy. In prior research, I found that mandatory generation divestiture (or not) had a significant impact on IOUs' policy preferences when it came to subsequent climate policy proposals. Specifically, in California and Massachusetts, states that required generation divestiture, IOUs were neutral to supportive of those states' economy-wide greenhouse gas (GHG) policies whereas, in Oregon, which allowed IOUs to continue to own generation, IOUs opposed the policy and worked successfully to weaken it (Basseches 2024).

Figure 1.3 shows the states that required generation divestiture for incumbent IOUs. It is worth noting that most, though not all, states with retail competition required generation divestiture. It is also worth noting

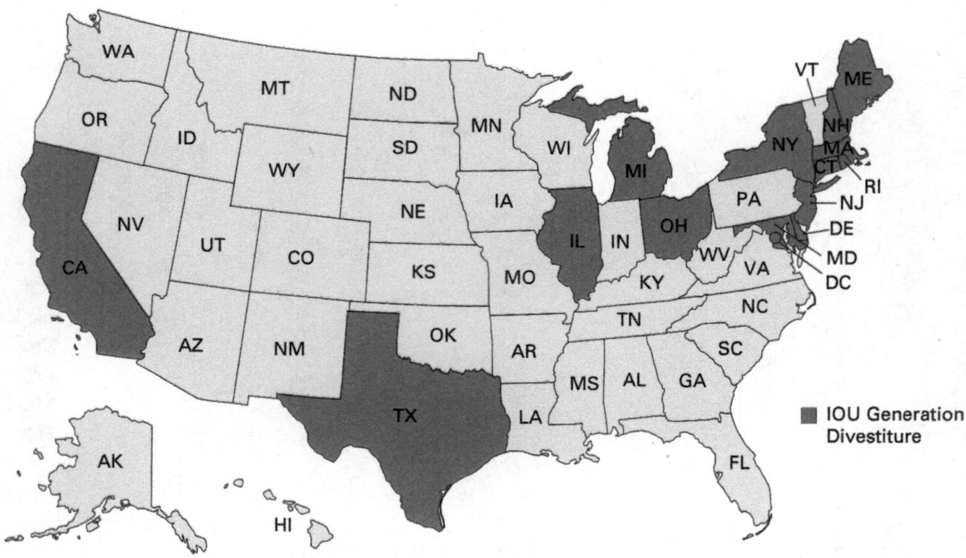

Figure 1.3
States with mandatory generation divestiture for IOUs. *Sources*: Harvard Electricity Policy Group via 6/28/22 Brattle Group Presentation to South Carolina Electricity Market Reform Measures Study Committee, Author's Research on Restructuring Statutes.

that plenty of states with at least some degree of wholesale competition—states that are part of ISOs/RTOs—still allow IOUs to own generation and do not allow retail competition; Oklahoma is an example. We will see how all these distinctions matter for the political economy of RPS policy design in the chapters that follow. But first, let us briefly review some of the key federal and state-level policy developments that helped liberalize the electricity system to result in the hodgepodge of state-level electricity policy regimes that one finds throughout the country today.

Our story begins with the Federal Power Act (FPA) of 1920, which was the first major federal law affecting the IOU industry and was the statute that created the FERC.[7] What is especially noteworthy about the FPA is that it was limited in the authority it vested in the federal government; instead, congressional policymakers made the deliberate choice to leave the states in the driver's seat when it came to electricity regulation (Boyd and Carlson 2016). Fifteen years later, the PUHCA of 1935 marked the federal government's first effort to rein in the corporate empires that Insull and his cohorts had created.[8] The PUHCA "required utilities to operate in contiguous geographic service territories, mandated robust disclosure to the Securities and Exchange Commission (SEC), and imposed operational limits on the types of business activities utilities engaged in" (Kovvali and Macey 2022, 4). However, key parts of this law were repealed by the Energy Policy Act of 2005, signed into law by George W. Bush, and IOUs found ways to skirt its purpose both after the 2005 repeal and in the intervening years, even before it (Hempling 2020; Kovvali and Macey 2022).

When it comes to the era of restructuring, the first important federal law was the Public Utility Regulatory Policies Act (PURPA) of 1978. PURPA came about in the aftermath of the Arab oil embargo, at a time when energy security and energy independence were paramount concerns to federal policymakers in Washington, DC. PURPA did a number of important things, but the most important for our purpose was the designation of "qualifying facilities" (QFs) and the associated requirements PURPA imposed on vertically integrated IOUs to purchase power from those QFs at the "avoided cost" rate. Critically, the definition of a QF included that its ownership had to be at least 50 percent independent of IOUs or their holding companies. The "avoided cost" provision meant that if a QF could provide electricity less expensively for ratepayers than it would cost for an IOU to generate that same volume of electricity by building a new power plant (i.e., at the avoided cost rate), then the IOU was required by law to interconnect the QF to its system and purchase its power. Of course, the cost to the IOU of the QF's power would be entirely passed through to ratepayers, but the IOU would not collect an additional return from ratepayers as it would if the IOU had built and owned the power plant itself (Boyd 2013; Isser 2015).

PURPA's QF provisions marked the first and perhaps only significant curtailment of IOUs' monopoly power by the US Congress since the PUHCA, and PURPA also gave birth to an entire new industry within the electricity sector, which would offer some measure of competition with IOU-owned generation (however, as we will see, how much actual competition there was would still vary by state). These new companies that emerged became known as independent power producers (IPPs). What is noteworthy about IPPs is not only their ability to offer electricity at a lower cost to ratepayers, and their relative independence from monopoly IOUs,[9] but also their eventual ties to the renewable energy industry—an industry that was still nascent back then, but that would go on to become a significant part of the US electricity generation mix and would be critical in the fight against climate change (Stokes 2020). By 2013, about 85 percent of total utility-scale renewable electricity generation in the United States was owned by competitive IPPs as opposed to monopoly IOUs (Kelsey and Meckling 2018).

However, the growth of IPPs did not happen overnight in the aftermath of PURPA. Indeed, it was quite slow at first. PURPA may have opened the door, but hardly the floodgates. Even by 1992—nearly a decade and a half after PURPA—competitive IPPs generated only 6 percent of electricity consumed in the United States; the lion's share still came from vertically integrated IOUs (Isser 2015).

IPPs gained much more of a foothold in the generation business in the late 1990s and into the early to mid-2000s, thanks to the federal Energy Policy Act of 1992, and even more importantly, thanks to a wave of state-level restructuring laws. The Energy Policy Act of 1992 included provisions that curtailed *a few* of the uncompetitive advantages IOUs still enjoyed vis-à-vis IPPs in the electricity marketplace and gave rise to the FERC orders discussed previously, authorizing the creation of ISOs/RTOs among states and market participants who were willing to agree to one (Pub. L. 102–486). Yet it was the *state-level* restructuring laws that followed, beginning in the mid-1990s, that created today's landscape, in which IOUs still dominate the generation business in the Southeast and large parts of the West, but in the rest of the country, IPPs enjoy a significant share of the generation business, and in some states, all of it. Table 1.3 shows the distribution in percentages of IPP-owned generation in 2021, with our seven case study states bolded.[10]

State-level restructuring laws swept through the country beginning with California in 1996, with a total of twenty-three states having adopted one between 1996–2000 (Isser 2015; Lee 2020). The precise number of restructured states that exists today is something of a moving target, as bills continue to be introduced in a number of states, as recently as 2023 (Journal Record Staff 2023). Even South Carolina, in the heart of the vertically integrated Southeast region, recently convened a legislative committee that

Table 1.3
Percentage of electricity generated from independent power producers (IPPs) in 2021, by state

Range	States	
0–20 percent (21 states)	Alaska	Utah
	Kentucky	Georgia
	Tennessee	Iowa
	South Carolina	North Carolina
	Florida	Hawaii
	Louisiana	Wisconsin
	Mississippi	Michigan
	Arkansas	Virginia
	Missouri	Alabama
	Wyoming	Arizona
	Washington	
21–40 percent (10 states)	North Dakota	West Virginia
	Indiana	Colorado
	Oregon	Nevada
	Minnesota	Idaho
	Nebraska	New Mexico
41–60 percent (7 states)	Kansas	Montana
	South Dakota	Vermont
	Oklahoma	New York
	California	
61–80 percent (2 states)	***Texas***	Delaware
81–100 percent (10 states)	Maine	***Pennsylvania***
	Maryland	Illinois
	Massachusetts	Connecticut
	Ohio	Rhode Island
	New Jersey	New Hampshire

Source: US Energy Information Administration.

commissioned a report exploring whether that state should pursue restructuring reforms (Tsoukalis et al. 2023), though the chances of successful restructuring legislation in that state seem low at the time of this writing given the political power of that state's IOUs (Basseches 2024; Harrison and Welton 2021).

Figure 1.4 shows a timeline of the key policy developments at both the federal level (in brown boxes) and at the state level (in green boxes), in the case study states in this book, that have just been described. The electric restructuring era is characterized by the increased policy activity in the 1990s and early 2000s. Of particular note is that Oklahoma adopted a partial restructuring law in 1997 (Oklahoma Senate Bill 500 1997), but it was never implemented, since Senate Bill 220, which would have been required to effectuate the 1997 Act, died in a legislative conference committee (Oklahoma Electric Restructuring Advisory Committee 2001). Oregon

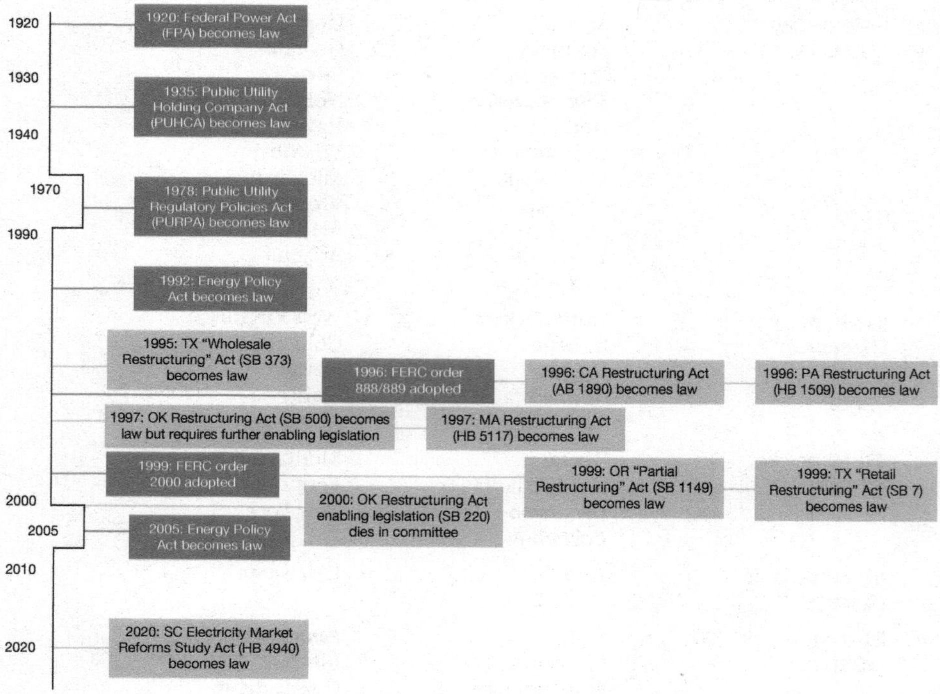

Figure 1.4
Policy timeline of key events in era of electricity restructuring. *Source*: Author.

adopted only a "partial restructuring" law in 1999. Critically, while Oregon adopted *partial* retail choice, it allowed incumbent IOUs to remain in the generation business, which my previous research found was a key factor in IOUs' opposition to climate policies in Oregon that their counterparts were neutral-to-supportive of in fully restructured states like California and Massachusetts (Basseches 2024).

How the IOU Business Model Shapes State-Level Policy Preferences

Because of Insull's regulatory compact described earlier, regulated-monopoly IOUs make money for their shareholders in a fundamentally different way than competitive businesses, including those in the energy sector. This, in turn, affects their policy preferences when it comes to RPS policies in ways that will become evident throughout this book. Entire bodies of legal scholarship debate the particularities of utility rate regulation, but at a rudimentary level, IOUs pass through the costs of operating the parts of the electricity system they own directly to their captive ratepayers. They neither profit from, nor lose money on, operating the assets they own—whether generation plants, transmission, or distribution infrastructure. This includes fuel costs, in the case of fossil-fueled generation such as coal-fired or gas-fired power plants; recall the quote at the beginning of this chapter that "fuel is free." Instead, these companies profit by expending capital on *building* the physical assets (whether generation, transmission, or distribution assets) and petitioning the state PUC to allow the maximum rate of return (as a percentage of those costs expended) to be collected from captive ratepayers, on top of the operating costs (Isser 2015).

What this unique business model means in terms of their RPS policy preferences is not always straightforward. When confronted with the prospect of a state-level mandate to procure electricity from a certain type of generation technology (i.e., an RPS), a generation-owning IOU must ask itself whether it can make more money for its shareholders by complying with such a mandate or by building whatever generation assets can earn them the highest return, unconstrained by de facto subsidies for some technologies but not others. By contrast, the calculation for an IOU that owns only transmission and distribution assets, but no generation assets, is quite different. In such cases, the IOU must only determine whether the RPS will increase the demand for new transmission and distribution infrastructure, which it can build and capitalize on. If the answer is yes, or even maybe, support for the RPS may be in the IOU shareholders' best interest, and thus, the IOU may lobby accordingly.

On the other hand, for a competitive IPP, the calculation about how to lobby strategically on an RPS proposal is entirely different. Simply put, if the IPP is in the business of owning and operating generation assets using green technologies that the RPS will reward, the decision to support is an easy and emphatic "yes." If, on the other hand, the IPP sees its future in maintaining the fossil-fueled status quo, by predominantly owning and operating fossil-fueled generation plants that the RPS seeks to penalize, then the opposite position on such policy proposals can be expected.

Other Political Actors in State-Level Renewable Electricity Policymaking

Aside from, of course, the state-level policymakers whose collective decisions form the outcome variable that this book seeks to explain (namely, variation in the design of seven states' RPS policies), several other important types of political actors constitute the cast of characters that will appear in the rest of this book. The most important are the regulated-monopoly IOUs, since it is *their* policy preferences, driven by the structure of the electric utility sector, that I argue is consistently the most helpful in explaining RPS policy design variation *beyond* those factors the literature has already emphasized. Readers will find a list of IOUs for the seven case study states in table 1.1. However, several other key actors merit mention as well.

IOUs are not the only providers of electricity. POUs, which can take the form of municipally owned utilities (e.g., Los Angeles Department of Water and Power, Austin Energy, Eugene Water and Electric Board) or public utility divisions (e.g., Sacramento Municipal Utility District, Lower Colorado River Authority, Columbia River People's Utility District), provide electricity to customers within their geographically defined service territories as well.[11] A third type of electricity provider, besides IOUs and POUs, is rural electric cooperatives (e.g., KAMO Electric Cooperative, Mid-South Electric Cooperative Association, Midstate Electric Cooperative). These are private, but not-for-profit, membership-based utilities that tend to serve geographic areas where it would be uneconomic for IOUs to set up shop, given how sparsely populated they are, with so few customers per square mile. Nationwide, IOUs serve about three-quarters of US electricity customers with the remainder served by POUs or rural cooperatives.

There are three different classes of electricity customers, each of which is defined in law and receives its own policy treatment by FERC and by state PUCs: industrial, commercial, and residential. Industrial customers are large, energy-intensive firms often engaged in manufacturing or other industrial processes (e.g., chemical plants, steel plants, oil refineries). Commercial customers are generally businesses of varying sizes (e.g., hotels, office buildings, restaurants), and, finally, residential customers are households (Environmental Protection Agency 2023). Clearly, commercial and industrial customers tend to be business interests, or private actors, whereas residential customers tend to be public interests, or public actors.

Each of these customer classes can be tied to a set of interest groups that lobby in state capitals, seeking to influence the decision-making of policymakers—whether elected legislators or appointed or elected (depending on the state) PUC commissioners (Heern 2023)—in ways that serve their interests. Typically, when it comes to electricity policy, these interests include lower rates, more reliable service, and, to varying degrees, perhaps environmental concerns as well. Table 1.4 provides examples of just a few of the interest groups encountered in my archival and interview-based research that were associated with each of the three customer classes.

Another type of political actor that frequently lobbies state-level policymakers on electricity policy, such as RPS policies, is environmental groups.

Table 1.4
Selected examples of interest groups representing different customer classes

Industrial	Commercial	Residential
Industrial Customers of Northwest Utilities	South Carolina Small Business Chamber of Commerce	The Utility Reform Network
		Public Citizen
Texas Coalition for Competitive Energy	Pennsylvania Office of Small Business Advocate	Citizens Utility Board
		Consumers Union
Oklahoma Industrial Energy Consumers	The Energy Consortium	AARP
Industrial Energy Consumers of Pennsylvania		Citizen Power, Inc.
California Manufacturers and Technology Association		

These are not-for-profit, public interest groups that are concerned first and foremost with the environment. Of course, there is overlap between environmental groups and residential ratepayer advocates, especially in states with high levels of grassroots support for climate policy. Furthermore, over the years, changes in the energy economy driven by policies globally have increasingly brought cost concerns and environmental concerns into closer alignment; cleaner energy *is* often cheaper energy. Many of those I interviewed for this book—regardless of whether they lobbied on behalf of residential ratepayers or environmental priorities—said, not surprisingly, that they believed pro-consumer policy *is* pro-environment policy (and vice versa). Yet environmental groups of the sort listed in the first column of table 1.5 tended to have a portfolio of state-level environmental priorities that included but also went well beyond clean electricity policy.

Of course, a wide range of business actors with an interest in these policies exists as well. Business concerns that are narrowly focused on the cost of electricity are often best represented by the kinds of groups listed in the first two columns of table 1.4, but there are also plenty of non-IOU business actors with a broader set of concerns around RPS policies, including but not limited to the cost of electricity. In the case of what I call here and elsewhere (see Basseches 2024) "green business" actors, they see enormous business potential in the adoption of environmentally strong state-level RPS policies, because they operate in industries—such as wind, solar, and battery storage—whose business model is aligned with fighting climate change. In short, they stand to *profit* from environmentally strong RPS policies. But they are different from both residential ratepayer advocates and nonprofit environmental groups in that the primary driver behind their lobbying in favor of environmentally strong RPS policy is financial self-interest rather than altruism (although, of course they would not put it that way; Basseches 2024). The middle column of table 1.5 provides some examples of such groups.

Another category of business actors, which I call "traditional business" (Basseches 2024), have policy preferences that are incompatible with environmentally strong RPS policies. There is a voluminous literature that highlights instances in which business opposes environmental policy (e.g., Brulle 2013; Kamieniecki 2006; Layzer 2007). Undoubtedly, this is oftentimes the case; however, climate change is one environmental problem for which certain business actors (i.e., "green business") can be, and are, part of the

Table 1.5
Selected examples of environmental groups, "green business," and "traditional business" actors

Environmental Groups	"Green Business" Actors	"Traditional Business" Actors
Natural Resources Defense Council (NRDC)	California Solar Energy Industries Association (CalSEIA)	***Fossil Fuel Industries:***
Environmental Defense Fund (EDF)	California Wind Energy Association (CalWEA)	California Independent Petroleum Association
Environment California	SunPower Corporation	Chevron
Sierra Club	California Energy Efficiency Council	Valero
Union of Concerned Scientists	Silicon Valley Leadership Council	New England Petroleum Council
Conservation Law Foundation	Center for Energy Efficiency and Renewable Technologies (CEERT)	Western States Petroleum Association (WSPA)
Environment Northeast	Cape Wind, LLC	Oklahoma Independent Petroleum Association
Oregon Environmental Council	Conservation Services Group	Chesapeake Energy
Environment Texas	New England Clean Energy Council (NECEC)	Continental Resources
Coastal Conservation League	Avangrid Renewables, LLC	Pennsylvania Coal Association
Southern Environmental Law Center	EDF Renewables	Pennsylvania Oil and Gas Association
Conservation Voters of South Carolina	Iberdrola Renewables	***Other (Adjacent Industries):***
Oklahoma Sustainability Network	Renewables Northwest Project	California Mining Association
Clean Air Council	The Wind Coalition	California Chamber of Commerce (CalChamber)
PennEnvironment	Texas Renewable Energy Industries Association (TREIA)	NAIOP Commercial Real Estate Development Association
Pennsylvania Environmental Council	Texas Solar Power Association	Associated Industries of Massachusetts (AIM)
Clean Water Action	Sunrun	Northwest Pulp and Paper Association
Action PA	Pine Gate Renewables	Oregon Cattlemen's Association
	South Carolina Solar Business Alliance	Northwest Food Processors Association (NWFPA)
	Southern Alliance for Clean Energy	Texas Association of Business
	Bergey Windpower	Texas Oil and Gas Association
	Oklahoma Wind Power Initiative	Pennsylvania Business Roundtable
	Oklahoma Renewable Energy Council	Pennsylvania Chamber of Business and Industry
	Standard Solar, Inc.	Pennsylvania Manufacturers Association

solution, as well. As Mildenberger (2020) points out, both business and labor are divided into pro-climate and anti-climate coalitions. Others have argued similarly (e.g., Downie 2019; Meckling 2011, 2015, 2019; Stokes 2020).

Nevertheless, there are certainly powerful and entrenched business actors—most notably the fossil fuel industry—whose business model is incompatible with climate protectionism, and which can therefore be expected to oppose environmentally strong RPS policies. Furthermore, scholars have documented an asymmetry in the political power of these "traditional business" actors relative to "green business" actors (Brulle and Downie 2022; Stokes 2020). As Mildenberger (2020) puts it, the former enjoy "double representation." While the coal, oil, and gas industries are the most obvious examples of "traditional business" actors, they are not the only ones, and their political activity is not limited just to major energy-producing states.

Many energy-intensive and/or trade-exposed industries outside of the electricity sector whose business models still depend to some extent on the availability of fossil-fueled energy, as well as broad-based trade associations whose largest members (and most significant funders) are oftentimes fossil fuel companies, can be considered part of the "traditional business" category based on the positions they take on renewable energy policy (Brulle and Downie 2022; Hall, Culhane and Roberts 2024). These types of business actors, both inside of and adjacent to the fossil fuel industry, are exemplified in the righthand column of table 1.5. To the degree government subsidies and private sector finance shifts over time in favor of "green business" actors (as we began to see since the passage of the Inflation Reduction Act in the United States, but this progress is under constant threat as we are now seeing in the second Trump administration), the balance between "green business" and "traditional business" actors *should* experience a corresponding shift in favor of the former, or at least so we would hope.

Finally, the era of restructuring created a new type of state-level political actor: retail power marketers (RPMs). These are corporate actors that are in the business not of generating or transmitting electricity but of *marketing it* to consumers in states that allow retail choice (see figure 1.4). The most famous—or infamous—of such companies was Enron. Today, a range of competitive RPMs are in the business of offering alternatives to the default service plans offered by incumbent regulated-monopoly IOUs. Sometimes, these alternatives come with promises of being less expensive or of offering customers more flexible rates; other times, they offer environmentally conscientious customers the opportunity to pay a bit *more* than the default service provider rate to "go green," promising electricity sourced exclusively from renewable generation. In states that already have retail choice, RPMs have become influential political actors; in states that do not, they are often the instigators at the center of efforts to lobby for further restructuring to promote competition and expand the market for their industry into new states.

This chapter has provided key background information on a complex and highly technical area of public policy (electricity), with a wide range of stakeholders that regularly attempt to shape it. Historically, most of the action in this policy area has been in US state capitals rather than in Washington, DC. Given the significant implications of electricity policy for climate policy, and the fact that the centrality of electricity policy will only grow to the extent the nation pursues a climate strategy of "electrifying everything," our knowledge of the political economy of electricity is underdeveloped and requires a great deal of further elaboration.

This book begins to offer that, focusing on the implications for state-level RPS policies, which have garnered significant scholarly attention as the most prominent example of state-level renewable energy policy diffusion (e.g., Carley et al. 2018; Lyon 2016; Parinandi 2023). Yet this extant literature, while robust and informative, does not emphasize the crucial ways in which the political economy of electricity affects the policy preferences of corporate and other political actors who then lobby policymakers, with varying degrees of success, to enshrine their preferences into law and regulations, exemplifying the notion of "policy as prize" (Hacker and Pierson 2014, 648).

In the next chapter, I elaborate the research questions and theoretical frameworks that guide the investigations that follow. I also detail the rationale informing my case selection and the methodological approach I use to answer my questions. I review literatures associated with state-level interest group politics, business–government relations, American political economy, and policy design and policy feedback effects. I highlight what is useful from each of these literatures in understanding divergent RPS policy outcomes, and I also argue that a more complete understanding requires appreciating the importance of electric sector structure in shaping IOUs' preferences.

2 The Puzzle of Renewable Energy Policy Design

The importance of "policy design" was brought to the fore in Schneider and Ingram (1997)'s *Policy Design for Democracy*, among other places. At the core of this book is the idea that attention to policy designs, which are "observable phenomena found in statutes, administrative guidelines, [etc.]" (Schneider and Ingram 1997, 2), should be a requirement for any theory of public policymaking, or of political influence in the policymaking process. This is because policy designs have a "dramatic effect on the distribution of wealth and other resources within society" (Schneider and Ingram 1997, 101).

Unfortunately, a careful focus on policy design has been lost in much of the literature of the years since, but a renewed focus on the study of "American political economy," in the tradition of political scientist E. E. Schattschneider, offers a tremendous opportunity to revive it (Hacker et al. 2021). Schattschneider's work emphasized the idea that "policies make politics" (Schattschneider 1935), meaning that there is an iterative relationship between organized political power and the design of public policy that can be used to reify or disrupt political power imbalances within the broader society.

Scholars have long been interested in this relationship between political power and public policy, and, in particular, in the degree to which private interest actors exert disproportionate influence over public policymaking (Chen 2019). This question has been asked at the federal level of the US government in general (e.g., Baumgartner et al. 2009; Gilens and Page 2014) and with regard to environmental policy in particular (Kamieniecki 2006). The findings of this literature on business power are partial and inconclusive. As Chen (2019) writes, "Looking broadly over the literature, there is still no easy consensus about business influence and whether it distorts

American politics and policy. . . . One could be forgiven for thinking that meaningful generalizations are still not possible" (Chen 2019, 34).

Regarding environmental policy at the federal level across all three branches of government, looking systematically, Kamieniecki (2006) finds that business actors on the whole are not necessarily more influential than nonbusiness actors—at least not in readily observable ways. Yet one of the most interesting findings in Kamieniecki's study is just how infrequently business actors took a position on the record, especially when it came to legislation in Congress: "Business interests do not take a position on almost 80 percent of the bills considered by Congress within this policy domain . . . Perhaps equally remarkable is that the business community, either as a whole or in part, opposes legislation in relatively few instances" (Kamieniecki 2006, 88).

As Kamieniecki suggests, it is possible that business actors are more influential than the congressional record would reflect. He writes, "Perhaps anticipating objections from congressional supporters of business, sponsors of environmental bills probably draft legislation with the interests of business in mind . . . It is also possible that business interests had rejected earlier proposals in private communications but were willing to embrace the revised legislation because objectionable provisions were eliminated and recommended changes were made" (Kaminiecki 2006, 88).

This suggestion on Kaminiecki's part seems consistent with what Page, Seawright, and Lacombe (2019) call "stealth politics" as well as what Broockman (2012) refers to as the "problem of preferences." As highly sophisticated political actors, business actors tend to know better than to announce their policy preferences publicly, but this does not mean that they do not hold such preferences. At the same time, we cannot presume from their silence that they agree. We know that when it comes to climate and renewable energy policy in particular, business actors often have fragmented or even competing preferences, depending on whether they stand to be winners or losers from the clean energy transition and the policies that advance it (Downie 2019; Meckling 2011, 2015; Mildenberger 2020).

We also know that at least certain business actors wield significant political power at the *state* level (Grumbach 2022; Hertel-Fernandez 2019), and that, to date, climate and renewable energy policymaking has been primarily the province of state governments (Basseches et al. 2022; Carley 2011; Rabe 2004). Yet, with the important exception of Stokes (2020), there has been minimal empirical work that evaluates the influence of *particular* business actors in state-level renewable energy policymaking. As we will see, to understand variation in state-level RPS policy design along *two* dimensions—both environmental strength *and* distributive politics, or the question of "who pays" for environmentally strong policy (Basseches 2024; Mildenberger 2020)—it is helpful to separate out business actors' preferences at an even more granular level. By taking the long view, through longitudinal historical case studies, we can also see how policy feedback effects from the era of restructuring and electricity market liberalization shaped the subsequent preferences of IOUs when it came to RPS policy design.

State-Level RPS Policies as a Tool to Fight Climate Change

The first thing to know about US climate policy is that it has long been a "bottom-up" endeavor, led by state governments, not Washington, DC (e.g., Parinandi 2023; Rabe 2004; Stokes 2020). This is true as ever at the time of this writing, during the second Trump administration. Of course, there is a certain irony to this fact, given that climate change is a problem of *global* proportions, but such is the "double-edged sword" of American climate and energy federalism (Karapin 2020). In fact, until the 2022 enactment of the federal Inflation Reduction Act (IRA), the US Congress had never passed even a single piece of legislation explicitly addressing climate change,[1] instead pursuing a strategy of piecemeal "institutional layering" (Mildenberger 2021). Even with the IRA's passage, historic and important as it is, the lion's share of government implementation effort falls to the states, specifically state energy offices and state PUCs (Ciulla et al. 2022).

During the decades-long period in which states acted alone on climate, RPS policies emerged as the single most prevalent in a suite of state-level climate policy instruments, and consequently, a growing academic literature emerged around these RPS policies (e.g., Bromley-Trujillo et al. 2016; Carley 2011; Carley et al. 2018; Fischlein and Smith 2013; Parinandi 2023; Yin and Powers 2010). So, what is an RPS policy, which states have adopted one, and how do they vary from one state to the next?

How RPS Policies Work: Policy Design Is Key and Variable

RPS policies are targeted at the electricity sector (as opposed to other carbon-intensive sectors, such as transportation, buildings, industry, etc.), which represents approximately one-quarter of US greenhouse gas (GHG) emissions at the time of this writing. However, given the strategy of "electrify everything" (i.e., other sectors) as a means of decreasing overall GHG emissions economy-wide (Harper et al. 2023), one could argue electricity is the most urgent source of emissions to tackle. At the time of this writing, efforts to reduce electricity sector emissions are under threat by the repeal of key IRA incentives and by surging load growth, due in large part to artificial intelligence (AI), and the electricity-ingesting data centers that power it.

Approximately 58 percent of total US electricity sales are governed by an RPS policy. The remainder either occur in states without an RPS or from electric utilities not subject to the RPS, which tend to be rural electric cooperatives and, in some states, POUs (Barbose 2021). About half of all growth in US renewable electricity capacity and generation since the year 2000 has been attributed to state-level RPS policies (Barbose 2023).

At the most basic level, RPS policies stipulate that, by specified deadlines, specified percentages of electricity generated or consumed within a particular state must come from renewable sources. That is where their commonality across states starts and ends. And indeed, even the "renewable sources" part varies, depending on what state law stipulates regarding resource eligibility. For example, Pennsylvania's RPS policy—admittedly, it is termed locally (and more accurately), the Alternative Energy Portfolio Standard (AEPS)—considers resources like "waste coal," a by-product of traditional coal-fired electricity production that can also be used to generate modest amounts of electricity, to be eligible for compliance with the standard (Glenna and Thomas 2010; Rabe and Mundo 2007). As Barbose (2021, 3) points out, RPS policies are "never designed the same way in any two states."

Not only do RPS policies vary when it comes to the stringency of their targets and timelines (Carley and Miller 2012), the aspect of their design that tends to get the most attention, they also vary on many other important dimensions. The most important of these is whether they are mandatory or voluntary; in other words, does the electric utility or any other actor with compliance obligations under the RPS face any penalty or consequence for failure to meet the prescribed standard? As of 2021, some thirty-one US states had adopted a mandatory RPS policy while an additional six had adopted only voluntary or aspirational targets (Barbose 2021; Carley et al. 2018; Vasseur 2014).

Another important way RPS policies vary by state is when it comes to exactly which actors are regulated. Electric distribution companies (IOUs) are almost universally covered under the policies, but rural electric cooperatives and POUs are covered to varying (and generally lesser) degrees (Lyon and Yin 2010; Rountree 2019). Every state has specific coverage requirements that differ from one another (Fischlein and Smith 2013).

Another very important dimension of variation is resource eligibility; that is, which technologies count as "renewable" for the purpose of RPS compliance? Certain technologies—for instance, wind, solar, geothermal—tend to be universally embraced by environmentalists and are noncontroversially counted as "renewable" in most RPS statutes. However, other technologies, such as biomass, hydropower, landfill methane gas, and, most infamously, waste coal, are more controversial, and different state RPS laws treat them differently. Similarly, energy efficiency—electrons *not* delivered due to effective strategies to reduce demand known as demand-side management (DSM)—receives RPS credit in some states but not others. Many states divide technologies into multiple "tiers" or "buckets," valuing the resource or technology differently depending on the tier/bucket in which it appears. A classic example of this is the "solar carve-out," where a designated percentage of generation must come from solar specifically; this was part of an effort to boost solar photovoltaic (PV) technology, which was significantly more expensive than wind at the time that many of these RPS laws were first put in place (Fischlein and Smith 2013; Glenna and Thomas 2010).

Another feature of differentiation in RPS policy design is criteria related to the eligibility of the *facility* (as opposed to the technology) where electricity is generated. Eligible facilities are often defined by the date in which they entered into operation. Adhering to the principle of "additionality" (Raymond 2010), which some but not all RPS statutes do, would require that only facilities that commenced operation *after* the RPS policy's adoption be eligible. Facility eligibility can also be defined geographically, and depend, for instance, on whether the generating facility is physically located within the state's borders (Fischlein and Smith 2013).

RPS policies also vary regarding their rules around calculating compliance, which is usually done through the accounting system of renewable energy credits (RECs). The general rule of thumb is that one megawatt hour (MWh) of electricity generated produces one REC, which is essentially a certificate validating the unit of electricity's renewable attributes. Regulated actors such as IOUs need to produce a certain number of RECs at specified time intervals to demonstrate to regulators that they are complying with the policy.

Yet so many details of REC accounting vary, too, from one state to the next. Some states award a REC per unit *generated*, while others award credit only when the unit is actually *delivered*, thereby incentivizing adequate transmission capacity to get the electrons from where they are generated to where they will be consumed. There are geographic requirements that vary by state, which govern whether a REC must be generated/delivered within the state's geographic borders, anywhere within the applicable (often interstate) ISO or RTO, or, in at least one state's case, within the service territory of the specific utility. States can either fully allow, allow but cap, or disallow what are known as "unbundled" RECs. These are RECs that can be bought/sold/traded separately from the underlying, physical electrons that they represent, creating a liquid market for utilities to tap into flexibly, not unlike GHG allowances in a cap-and-trade system (Fischlein and Smith 2013; Raymond 2016). It is these rules regarding the geographic origins and destinations of RECs that likely explain a recent analysis' finding that state-level RPS policies can have "spillover effects" in neighboring states (Zhou et al. 2024).

Other accounting-related rules that differ from one state to the next include those that govern utilities' ability to comply more "flexibly." These can include proactive banking or retroactive borrowing of RECs for different periods of time—what Fischlein and Smith (2013, 291) call the "shelf life" of the REC. These rules can also resolve questions such as who owns the title to the REC once generated? They can include "credit multipliers," and rules related to the interaction of RPS-related RECs with "voluntary RECs" of the sort that many private companies use to demonstrate corporate social responsibility or that offer consumers options to voluntarily "go green" at the individual level (Fischlein and Smith 2013; Pennsylvania Public Utility Commission 2005).

Another important dimension of variation in the case of mandatory RPS policies as opposed to voluntary ones is, of course, what the penalty is for failing to comply with the RPS's requirements. Penalties usually take one of two forms. They can either be traditional fines in amounts stipulated by law or regulation, or they can be alternative compliance payments (ACPs). The idea in the case of the latter is that, for every obligatory REC that a regulated actor falls short of producing in a given compliance period, that regulated actor may instead "purchase" a nonexistent REC by making an ACP, which, similar to a fine, can be set at different levels by statute or regulation (Fischlein and Smith 2013). And finally, state RPS policies can be characterized by any number of extraneous provisions, distinctive to each state, that tend to be highly technical in nature and tend to have the effect of shifting financial costs or risks associated with RPS compliance from one actor to another. These can include provisions known to insiders as "off-ramps," "safety valves," "cost caps," or "automatic adjustment clauses," to name just a few (Basseches 2024).

This book will give special attention to these extraneous provisions because they are typically the direct result of interest group lobbying. They are also typically preconditions that powerful political actors such as IOUs demand from policymakers in exchange for their support of the overall RPS. They are, for instance, included among the "five basic needs, [or] must-haves" that the Oregon IOU lobbyist the last chapter opened with explained were gotten in exchange for supporting that state's RPS legislation. As we will see, far more often than not, they have the effect of shifting these costs/risks away from IOUs and onto electricity ratepayers, including residential, commercial, and industrial classes.

The preceding paragraphs have covered a wide range of policy design features whose emergence merits explanation. Explaining them requires moving beyond considering RPS policy adoption as a binary variable. It also requires moving beyond focusing only on variation in RPS targets and timelines, despite these being the most salient and accessible features of state-level RPS policies. Recently, some studies have begun to more fully appreciate the extraordinary range of policy design features that characterize RPS policies (e.g., Carley et al. 2018; Fischlein and Smith 2013; Rountree 2019).

Perhaps the recent study that is the *most* attentive to this variability is Parinandi (2023), whose unit of analysis is at the RPS "subpolicy" level, and who compiles a dataset of discrete RPS "policy feature" adoptions across all thirty-seven states that adopted RPS policies between 1983–2011. As Parinandi correctly points out, "Imagine that I compare the RPS policy regimes of Michigan and Ohio but ignore adoption at the granular level of policy features and look only at whether Michigan and Ohio adopted RPS policy regimes. The RPS policy regimes of the two states would appear identical even though they differ substantially in terms of what sources and processes are considered to be 'renewable' and whether electric utility companies must meet carve-outs or technology minimums" (Parinandi 2023, 45–46).

Parinandi's excellent study is mostly quantitative and is concerned with policy *diffusion* across US states rather than *how interest group politics shape* which design features emerge or disappear in the first place. In contrast, Stokes (2020) centers interest group politics in explaining the retrenchment of four states' RPS and Net Energy Metering (NEM) policies. Yet Stokes is less focused on explaining the emergence or disappearance of particular RPS policy design features at the "subpolicy level," of the sort Parinandi analyzes.

This book will draw on the significant conceptual and methodological strengths of both Parinandi (2023) and Stokes (2020). Following Stokes, I will focus on interest group politics and on the relationship between interest group preferences and policy design. Following Parinandi, I will attend to particular "subpolicy features" to uncover interest group influence that may not be apparent if viewing the policy at a more macro level. I will draw on a variety of triangulated qualitative data sources to explain the interest group politics behind RPS policies in seven states, which together house one-third of the US population.

Before proceeding further, I must briefly discuss the *venues* in which state-level RPS policies are formulated. In such venues, interest groups contend for influence over policy design, viewing "policy as prize" (Hacker and Pierson 2014, 648). Here, too, it is useful to draw on Parinandi (2023), who emphasizes the dual roles of state legislatures and state PUCs in cocreating a state's RPS policy regime. For reasons Parinandi discusses, with a few exceptions, more of the designing of RPS policies happens in state legislatures, but PUCs nevertheless retain a vital role that goes far beyond passive implementation (Parinandi 2023). While the authority (and therefore ambition) of PUCs is constrained by state constitutions and legislative statutes, the hard work of fleshing out the finer details that the legislature is often silent on falls to these PUCs. These details can be critical in their distributive impacts, and therefore, to the policy preferences of the same interest groups that vie for influence in the legislature. Therefore, PUC rulemaking must be considered when assessing RPS policy outcomes, especially along multiple dimensions (Basseches 2024). Therefore, like Parinandi (2023), this book considers RPS policymaking activity in both venues.

It is also worth noting a third venue from which RPS policies can originate, and that is state-level ballot initiatives. Such were the origins of the RPS policies in states like Colorado and Washington (Parinandi 2023). This book will not analyze such cases, but they represent an important avenue for future research, given the dearth of studies assessing how their politics compare to those of legislatively and regulatorily created RPS regimes.

Profile of RPS-Adopting States

Thirty-seven US states have adopted *some form* of an RPS policy, though these include six states in which the policy is entirely voluntary, and others in which it has been allowed to expire (Barbose 2021; Carley et al. 2018; Parinandi 2023). Although voluntary RPS policies are obviously far inferior to mandatory ones in terms of their environmental strength, voluntary RPS policies nevertheless indicate an inclination on the part of a state's policymakers toward nontraditional sources of electricity generation (Vasseur 2014). Therefore, voluntary RPS policies are included in this book, and indeed represent one indicator of meaningful policy variation.

When one considers only *which* states have adopted an RPS to date, setting aside for the moment the strengthening or weakening of the policies over time, the map looks different than a scholar or observer of state-level climate policy might expect. Figure 2.1 shows the average partisan composition of the RPS-adopting state legislatures from 1995–2021, the period in which nearly all initial state RPS adoptions took place (Barbose 2023). As the figure shows, while all six states that adopted a purely voluntary RPS policy were "red states," plenty of red states—not only "blue states"—did adopt *mandatory* RPS policies as well. These include Texas and Pennsylvania, which this book will examine in depth. This pattern is interesting in light of the large, quantitative social science literature, which finds a strong, positive association between Democratic partisanship and/or liberal ideology and climate policy adoption (e.g., Bromley-Trujillo et al. 2016; Coley and Hess 2012; Haseloff 2024; Trachtman 2020; Vasseur 2014).

Another state-level characteristic that the literature finds to be associated with climate policy adoption is in-state fossil fuel production. It is intuitive that states that produce (and export) a great deal of coal, gas, and/or oil within their borders would be less inclined to adopt climate or renewable energy policy, since the fossil fuel industry would presumably be more indispensable to those states' political economies, employing more

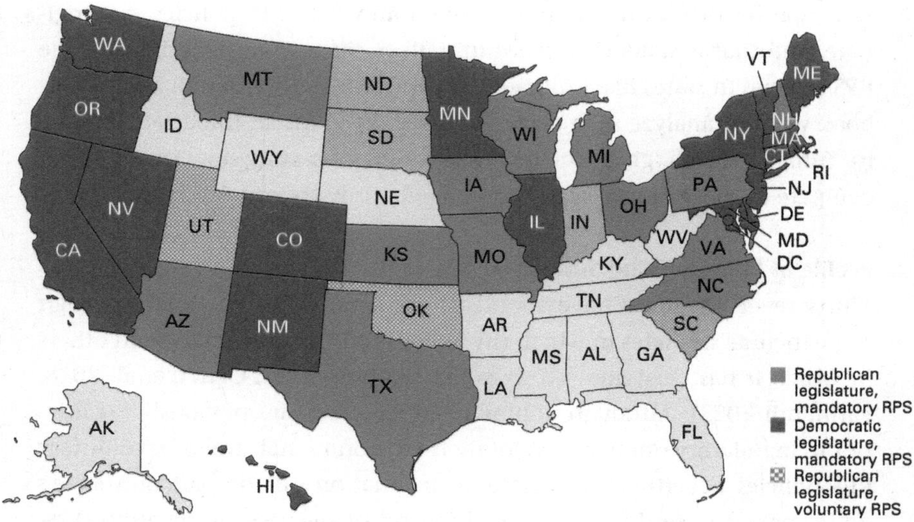

Figure 2.1
Average partisan composition of state legislatures of RPS-adopting states, 1995–2021.
Note: Source for partisan composition data is the Council of State Government's "Book of States."

residents (and voters) and constituting a larger share of the state's tax base. Those states would stand to be net economic losers of policies that would suppress demand for their product relative to states that produce less of those fuels, which might stand to be net economic winners, even irrespective of partisanship or political ideology (Bromley-Trujillo et al. 2016; Haseloff 2024; Huang et al. 2007; Thombs and Jorgenson 2020; Vachon and Metz 2006).

Figure 2.2 shows the relative average level of fossil fuel production of the RPS-adopting states during the same years. While many of the states that adopted mandatory RPS policies had relatively low levels of fossil fuel production (as we might expect), there are some notable and surprising exceptions, including California, Texas, and Pennsylvania—three of the seven states this book will examine in depth. Of course, these also happen to be very populous states, so their fossil fuel intensity would undoubtedly be diminished if measured on a per capita basis. Still, it is somewhat puzzling that a state like Texas or Pennsylvania—a "red state" with extremely high levels of fossil fuel production—would adopt a mandatory RPS *at all.*

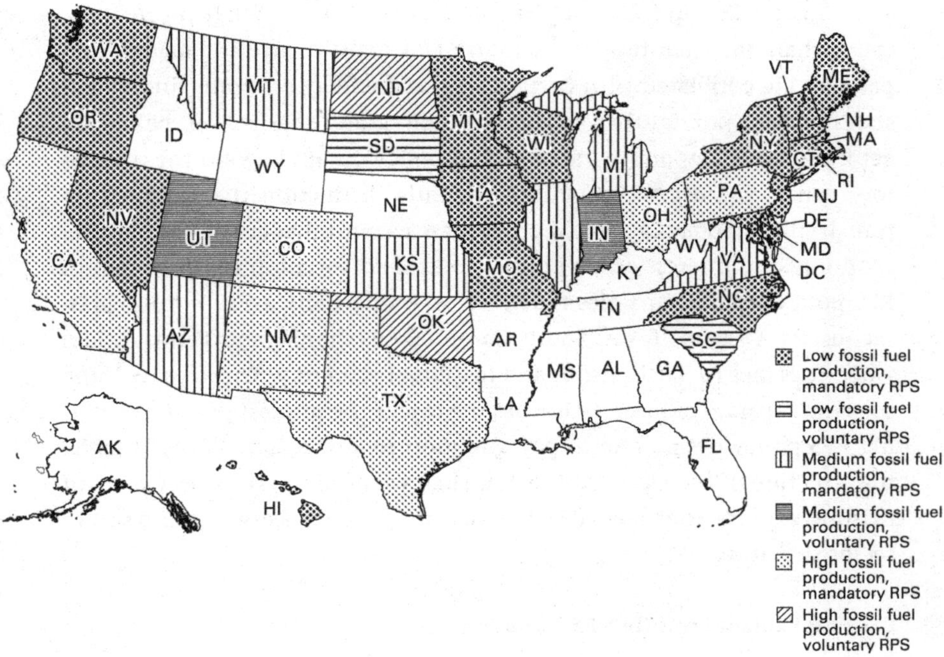

Figure 2.2

Average in-state fossil fuel (coal, gas, oil) production of RPS-adopting states, 1995–2021. *Note*: Source for fossil fuel production data is the US Energy Information Administration. "Low" fossil fuel production is defined as <100 trillion British Thermal Units (Btu); "medium" fossil fuel production is defined as 100–1,000 trillion Btu; "high" fossil fuel production is defined as >1,000 trillion Btu.

As we will see, Pennsylvania's RPS is environmentally weaker than many others', though puzzlingly it is still significantly stronger than South Carolina's, which is only *voluntary* or aspirational, despite South Carolina having far less fossil fuel production than Pennsylvania. Texas's *mandatory* RPS was among the very strongest RPSs in the country for at least the first decade that it was in effect (Carley et al. 2018), though as Stokes (2020) shows, Texas declined to strengthen its RPS policy in more recent years. By contrast, Oklahoma, whose fossil fuel production levels rival Texas's, adopted only a voluntary RPS, placing it among the very weakest in the country (though it is interesting, still, that Oklahoma took any RPS policy action at all).

As figures 2.1 and 2.2 suggest, there is a lot more going on in these states than the quantitative literature can explain. Even if macro-level patterns are correlated with partisanship and energy economy, individual states present puzzles that merit explanation, and that is even before we get into all the important variability in policy design beyond the mandatory versus voluntary distinction, the only distinction these maps capture. In the chapters to come, I provide an account that combines political economy and interest group politics to explain variation in the design of RPS policy regimes in states that vary as dramatically as California, Massachusetts, Oregon, Texas, South Carolina, Oklahoma, and Pennsylvania when it comes to partisanship and fossil fuel economy. The keys to doing so are to separate out the IOUs from the rest of the interest groups contesting RPS policy design and to pay attention to a third state-level variable: the structure of the electricity sector. First, we need to take better stock of existing explanations related to partisanship, energy economy, and utility sector structure.

Existing Explanations for RPS Variation

Partisanship

One of the strongest and most consistent predictors of variation in state-level climate and renewable energy policy is partisanship. In particular, climate policy adoption and diffusion tends to take place in jurisdictions that are under the control of Democratic politicians (Bromley-Trujillo et al. 2016; Coley and Hess 2012; Haseloff 2024; Trachtman 2020). At the level of the individual state legislator, Coley and Hess (2012) find that Republican legislators voted in favor of RPS bills much less frequently than Democrats did. At the jurisdictional level, Trachtman (2020) finds partisanship to be a crucial predictor as well, even more so when it comes to RPS policies than other climate policy instruments.

Partisanship has been found to inform not only climate and renewable energy policy *adoption* but also certain elements of policy *design*. Studies have shown that state government control by the Democratic Party was associated not only with overall RPS adoption but also with the distinction between mandatory and voluntary RPS policies. Of the states that adopted an RPS, Democrat-controlled states were more likely to have adopted a *mandatory* one (Haseloff 2024; Vasseur 2014).

In experimental research exploring how policy design of clean electricity standards, carbon taxes, and technology subsidy/investment-based policies affects public support, Bergquist, Mildenberger, and Stokes (2020) find stronger support among Democrats than Republicans when it comes to more environmentally stringent policy designs. At the same time, research evaluating climate policy stringency, both regarding RPS policies in particular (Carley et al. 2018) and state-level climate policy in general (Bergquist and Warshaw 2023), shows that despite the prevailing pattern of stringent climate policy regimes being associated with Democrat-controlled states, some Republican-controlled states have managed to adopt relatively stringent policies, or policy regimes, as well.

Qualitative work suggests a more nuanced picture of the role that partisanship may be playing at the level of actual policy-writing. In prior work, studying the bluest state in the country (Massachusetts), I found that differences in interest group politics, and particularly in the influence of environmental groups, explained why one climate bill was written to be much stronger than another, even though the partisan composition of the legislature was identical for both.[2] Stokes (2020) uses qualitative methods to study four states—Arizona, Kansas, Ohio, and Texas—that were Republican-led. While Stokes finds policy retrenchment, she also finds variability in initial policy strength as well as in policy durability; after all, some of these states needed to pass relatively strong policies initially for them to then be meaningfully retrenched. Furthermore, Stokes's story of retrenchment is more about interest group influence and political economy than it is necessarily about partisanship. As Stokes writes, "Battles over climate policy are fundamentally material: they are about who will get to own the assets of the energy system and the resulting profits" (2020, 4).

A recent essay by Egan and Mullin (2023) offers a clear-eyed view of the way in which partisan polarization has impeded state-level climate policy in the past but also suggests that there are reasons to question whether partisan polarization will be as much of an impediment in the future. One reason they point out that it may not be has to do with a shifting political economy, in which Republican-led states are often the ones seeing the greatest expansion of the clean energy economy, a trend accelerated by the IRA. This shift did not occur overnight, and so it is likely to have mattered to some degree when it came to the lobbying that took place in formulating state-level RPS policies in ways that might be obscured in quantitative studies that do not investigate determinants of variation in policy design at the "subpolicy level" (Parinandi 2023). Nevertheless, the weight of the literature is clear: Partisanship matters and, indeed, is often the single best predictor of climate and renewable energy policy adoption.

Energy Economy and Interest Groups

Studies have found that states with higher levels of fossil fuel production have been less likely to adopt robust climate and renewable energy policies (Bromley-Trujillo et al. 2016; Haseloff 2024; Thombs and Jorgenson 2020). Specifically, Bromley-Trujillo and coauthors find that a higher share of gross state product derived from mining is negatively associated with climate policy adoption (Bromley-Trujillo et al. 2016). Thombs and Jorgenson (2020) find that states with an RPS tend to have less fossil fuel production per capita than states without one. And Haseloff (2024) finds that states that are more economically dependent on oil and gas production specifically are less likely to have adopted a mandatory RPS. The notion that states whose energy economies are dominated by fossil fuel producers should be less likely to adopt renewable energy policy that directly threatens those fossil fuel producers has such strong intuitive appeal that it is routinely included as an explanatory variable in quantitative RPS studies (e.g. Huang et al. 2007; Matisoff 2008; Vachon and Menz 2006).

Thombs and Jorgenson (2020), in addition to finding in-state fossil fuel production per capita to be significant on its own, also demonstrate an interaction effect between in-state fossil fuel production and state government ideology. In states without significant fossil fuel production, they find greater likelihood that the state will adopt an RPS regardless of government ideology. By contrast, in major fossil fuel–producing states, they find that only liberal to moderate state governments are likely to adopt one. Thus, they conclude, fossil fuel industry presence plays its biggest role in impeding RPS policy in ideologically conservative states (Thombs and Jorgenson 2020).

Of course, a state's energy economy is also related to the landscape of interest groups that lobby its policymakers and how much influence such groups are likely to have. It should be noted, however, that while fossil fuel interest groups, for structural reasons, can be expected to be relatively stronger in states where such interests dominate the energy economy, these interest groups might be influential in a way that is disproportionate to their representation in a state's energy economy. Mildenberger's (2020) theory of "double representation" gives them more influence just by virtue of their benefitting from the status quo, and from veto points within existing political institutions, in a way that "green" groups do not. Other work suggests that interest groups opposing climate policies are disproportionately represented in the lobbying of trade associations (Brulle and Downie 2022).

In terms of interest groups *supporting* climate policies, Trachtman and Meckling (2022) examine state-level variation in the prevalence of pro-climate interest groups as a proportion of total energy-related interest groups. Interestingly, they find that such variation is closely correlated with variation in state partisanship. In "blue states," climate advocacy groups enjoy more representation, whereas in "red states," the opposite is true. This finding is consistent with more theoretical political science work about partisan-interest group alignment—namely, parties act as "teams" that recruit interest group support to uphold their institutional power (Bawn et al. 2012); clearly, the Democratic Party at the state level has recruited climate advocacy groups for its team. Moreover, clearly partisanship and energy economy do not always act independently from one another to explain climate and renewable energy policy outcomes. Interest group preferences and influence can play important mediating roles, and act as the mechanism through which either partisanship or energy economy might ultimately shape policy outcomes. Therefore, there is added value to qualitative work, like this book, which can parse how partisanship, energy economy, and other factors, too, combined to shape policy outcomes in particular cases. One such "other factor," to which we now turn, is utility sector structure.

Utility Sector Structure

This book is not the first to propose a connection between state-level utility sector structure and renewable energy policy. In an important study, Lee (2020) concludes that states that adopted restructuring laws were "fifteen to forty-three percentage points more likely to adopt a renewable portfolio standard and fifteen to thirty percentage points more likely to adopt a cap-and-trade program" than states that did not (Lee 2020, 1). This book lends further support to Lee's findings, but it also adds some important nuance and illuminates an important mechanism connecting state-level restructuring policies to subsequent state-level renewable energy policies.

First, the qualitative data mustered in this book illustrates in granular detail the interaction of IOU lobbying with the lobbying of other business actors, so that we can see exactly when, how, and why these actors' preferences converged or diverged to produce a range of RPS policy outcomes. Second, this book adds detail and nuance to the conceptualization of state-level restructuring policies. Just as no two RPS policies were designed in exactly the same way in the many states that adopted them, no two restructuring policies were designed in exactly the same way either. Careful analysis of policy texts reveals multidimensional differences impacting the subsequent landscape for RPS policymaking. Finally, this book identifies a mechanism connecting restructuring policies with subsequent RPS policies: policy feedback effects.

Policy Feedback Effects

Political scientists' concept of "policy feedback effects" is useful in adding a temporal element to the study of policy outcomes. Rather than seeing policymaking episodes as a snapshot in time, policy feedback researchers tend to see policymaking as a protracted contest between competing actors in which the outcomes of policies at time one alter the "terrain" on which actors compete to influence policy at time two (Hacker and Pierson 2014; Pierson 1993, 2000; Stokes 2020). Past policies can produce feedback effects related to "mass politics," which can impact voter-level political attitudes and/or behavior (Campbell 2012), or they can produce feedback effects that shape interest group preferences or power (Pierson 2000). The latter kind of policy feedback effects were found, by Stokes (2020), to play a significant role in the retrenchment of state-level climate and renewable energy policy. Stokes found that when states passed policies that failed to lock in pro-renewable shifts in distributive politics, often due to what Stokes calls the "fog of enactment," they were more easily retrenched (Stokes 2020).

This book will build on my previous work (see Basseches 2024) to show how state laws that restructured the electric utility sector created feedback effects on the interest group politics of state-level RPS policymaking. These effects were most pronounced when it came to one interest group actor: IOUs. As chapter 1 details, IOUs are not your typical business actor. Historically, they have been the most privileged private actors when it comes to electricity system governance. Social science research that demonstrates their immense, if not unmatched, political influence is rapidly accumulating (e.g., Basseches 2024; Culhane et al. 2021; Stokes 2020). Legal scholarships have found similar (Kovvali and Macey 2022, 2023; Peskoe 2023). This study is designed to be able to explore the interactions between partisanship, energy economy, and utility sector structure as they operate through the mechanism of interest group politics.

Research Questions, Data, and Methodological Approach

Outcome Variable: RPS Policy Environmental Strength

This study asks, first, what accounts for variation in the environmental strength of state-level RPS policy regimes, and second, what accounts for convergence in the distributive quality of environmentally moderate-to-strong RPS policies (Basseches 2024; Mildenberger 2020)? Regarding the first question, how can we measure variation in RPS's environmental strength? Researchers have proposed different ways.

One is Carley et al.'s (2018) measure, which includes a stringency component[3] as well as measures of a select number of other design features, such as resource eligibility, penalties, and geographical limits. Another is Fischlein and Smith's (2013), which is calculated based on the presence or absence of fourteen substantive provisions "believed to drive additional renewable energy deployment" (Fischlein and Smith 2013, 299). There is no perfect measure, but these two are both commendably sophisticated and nuanced. Since I began data collection for this project before Carley's measure was published and without knowledge of Fischlein and Smith's, I relied on a cruder measure as the basis of my case selection for which states' regimes I would study.

I used a combination of two factors—1) whether the state's policy was mandatory or voluntary, and 2) whether its nominal target was greater than or less than 50 percent—to categorize states' RPS policy regimes as "strong," "moderate," or "weak." Mandatory policies were considered "strong" if their nominal target was 50 percent or higher, and "moderate" if their nominal target was less than 50 percent and/or if they had been allowed to expire. All voluntary policies were considered "weak." Table 2.1 shows where the RPS-adopting states fall according to this categorization, with this book's seven case study states bolded.

As table 2.1 shows, the seven cases examined in this book vary across this categorical outcome variable, with three states (California, Massachusetts,

Table 2.1
State-level RPS policies categorized by environmental strength

Strong	Moderate	Weak (voluntary)
California	Arizona	Indiana
Hawaii	Colorado	**Oklahoma**
Illinois	Connecticut	North Dakota
Maine	Delaware	**South Carolina**
Maryland	Iowa	South Dakota
Massachusetts	Kansas	Utah
Minnesota	Michigan	
Nevada	Missouri	
New Jersey	Montana	
New Mexico	New Hampshire	
New York	North Carolina	
Oregon	Ohio	
Rhode Island	**Pennsylvania**	
Vermont	Washington	
Virginia	Wisconsin	
	Texas	

Sources: Barbose (2023), National Conference of State Legislatures (NCSL)

and Oregon) falling into the "strong" category, two states (Pennsylvania and Texas) falling into the "moderate" category, and two states (Oklahoma and South Carolina) falling into the "weak" category. Carley et al.'s (2018) measure was calculated for the period 1990–2014, while the period of my study is 1995–2021, but the three states I study that I classified as "strong" have the highest stringency scores of the seven according to their measure, followed by the two states I classified as "medium," followed by the two states I classified as "weak."[4] Fischlein and Smith (2013) examine only states with "mandatory" policies, but my "strong" states (California, Massachusetts, and Oregon) all perform better on their measure than my "moderate" states (Pennsylvania and Texas) do.

To provide further validation for my seven state case studies, we can look at the variability in the environmental strength of their climate policy regimes more broadly, including but not limited to their RPS policies, thanks to a comprehensive stringency score developed by Bergquist and Warshaw (2023). Bergquist and Warshaw's measure indexes twenty-five discrete state-level climate policies across a range of economic sectors at the unit of state-year. If we average the fifty states' scores in 2000, 2010, and 2020, we observe that California, Massachusetts, and Oregon ("strong" RPS policy states) are in the top quintile of all US states; Pennsylvania and Texas ("moderate" RPS policy states) are in the next quintile; and Oklahoma and South Carolina ("weak" RPS policy states) are in the quintile below that (Bergquist and Warshaw 2023).

Although clearly the environmental strength of RPS policies varies across states, when one reads the policy texts of the RPS policies in "strong" and "moderate" RPS states, one finds a similar pattern in the *distributive effects* of key provisions. For example, Senate Bill 1078, which created California's RPS, directed the California PUC to "allow an electrical company [IOU] to limit its annual [RPS] procurement obligation to the quantity of eligible renewable resources that can be procured with available supplemental energy payments,"[5] essentially guaranteeing IOUs access to ratepayer subsidies if renewables were to cost more than fossil-fueled generation. When California's RPS was updated in 2006, language was added directing the PUC to adopt "flexible rules for compliance,"[6] leading to a practice known as "earmarking" whereby IOUs could satisfy their compliance obligations ahead of time, despite insufficient transmission for that electricity to reach the grid. And then, when the RPS was once again updated in 2011, the California legislature included a series of "off-ramps" allowing IOUs a range of legally acceptable "excuses" for noncompliance based on contingencies that could have ended up making compliance more costly.[7]

In Massachusetts, a provision inserted into the initial 1997 RPS bill required the state to "revisit the feasibility" of the policy several years later.[8] Then, when the Massachusetts Department of Energy Resources (MDOER) wrote the regulations implementing the RPS, it established a novel (at the time) "unbundled" REC trading program that granted IOUs flexibility to meet the requirements by trading RECs in a liquid market with electric companies in other states rather than necessarily needing to build and connect new sources of renewable generation in Massachusetts.[9] In 2008, when the Massachusetts RPS was again revised, a provision appeared allowing Massachusetts IOUs to own and operate solar generation facilities rather than having to contract with IPPs to obtain solar electricity,[10] despite IOUs having been required to divest from generation during that state's restructuring period.

In Oregon, a "cost cap" was inserted into the 2007 RPS law, in §12: "Electric utilities [IOUs] are not required to comply with a renewable portfolio standard during a compliance year to the extent that the incremental cost of compliance . . . exceeds four percent of the utility's annual revenue requirement for the compliance year." In addition, an "automatic adjustment clause" provision was inserted, fundamentally altering the procedure through which IOUs recover costs from ratepayers, in favor of the IOUs. And in yet another provision, RECs were explicitly permitted to be "banked and carried forward indefinitely."[11]

In Texas, Senate Bill 7—which both completed the restructuring of that state's electric sector and created the initial RPS—also contained a provision that authorized the PUC to use ratepayer-funded incentives to encourage IOUs to procure renewable resources.[12] Then, when the Texas RPS target was increased in 2005, Senate Bill 20a included a provision stating that if the PUC orders IOUs to construct additional transmission-related facilities to bring additional renewable resources onto the grid, the PUC would be statutorily required to "find that the facilities . . . are prudent and *includable in the rate base*, regardless of the extent of the utility's actual use of the facilities [emphasis added]."[13]

Finally, in Pennsylvania, Senate Bill 1030, which created that state's RPS in 2004, included a "force majeure" provision that amounted to an "off-ramp" for IOUs, similar to what we saw in California. That provision states, "Upon its own initiative or upon a request of an electric distribution company [IOU] . . . the Pennsylvania Public Utility Commission, within 60 days, shall determine if alternative energy resources are reasonably available." If the PUC determines they are *not* reasonably available, the PUC "shall modify the underlying obligation . . . or recommend to the General Assembly that the underlying obligation be eliminated."[14] A subsequent section of the same bill contains an "automatic adjustment clause" similar to what we saw in the Oregon RPS legislation.[15] An "automatic adjustment clause" provides an expedited review of IOUs' requests for rate increases, not subject to the ordinary scrutiny and same opportunities for intervention by competing parties as they would be in an ordinary rate case proceeding (Basseches 2024).

Taken together, these provisions do not impact the environmental strength of the various states' RPS policies; instead, they shift financial costs/risks associated with relatively strong RPS policies away from IOU monopolies' shareholders and onto ratepayers. The degree to which the added burden is shared equally or unequally among the different classes of ratepayers, as we will see, varies by state, likely along with the quality of political representation (through lobbyists or otherwise) that these public and private actors enjoy in each state's political system.

Explanatory Variables: Partisanship, Energy Economy, Electric Sector Structure

Three major explanatory variables were considered in the case selection for this study, corresponding with the findings of the literature review earlier in this chapter. The first is the partisanship of the state legislature, as shown in figure 2.1. There are several ways I could have operationalized partisanship, but the one I chose for the sake of my initial case selection was to take an average of the percentage of the state legislature's membership belonging to each of the two major parties in each year under study, 1995–2021. If more often than not during this period the state legislature was controlled by the Republican Party, the state appears in red on the map, and if more often than not it was controlled by the Democratic Party, the state appears in blue on the map. As we will see in the chapters to come, there is important nuance in partisanship that figure 2.1 obscures (such as some blue and red states that are more "purple" than others, significant shifts in partisan composition over time, and the role of the governor's party), but given the qualitative and longitudinal nature of this study, I found the operationalization described appropriate for the purpose of initial case selection. Table 2.2 shows "red states" and "blue states" that adopted an RPS (irrespective of its strength), with the seven case study states again bolded. The second explanatory variable is the nature of the in-state energy economy, operationalized as levels of in-state coal, gas, and oil production, as shown in figure 2.2. Table 2.3 shows RPS-adopting states according to whether they produce "high," "medium," or "low" levels of fossil fuels within their borders, with the seven case study states bolded.

Finally, there is the structure of the electric sector, a variable that has been getting increasing attention lately in the renewable energy policy literature (e.g., Lee 2020; Parinandi 2023), but one that is difficult to operationalize due to the complexities and nuances discussed in chapter 1. For the purpose of case selection, I categorized this as a binary variable, with

Table 2.2
State-level RPS policies categorized by partisanship of state legislature

"Blue States"	"Red States"
California	Arizona
Colorado	Indiana
Connecticut	Iowa
Delaware	Kansas
Hawaii	Michigan
Illinois	Missouri
Maine	Montana
Maryland	New Hampshire
Massachusetts	North Carolina
Minnesota	North Dakota
Nevada	Ohio
New Jersey	**Oklahoma**
New Mexico	**Pennsylvania**
New York	**South Carolina**
Oregon	South Dakota
Rhode Island	**Texas**
Vermont	Utah
Washington	Virginia
	Wisconsin

the key distinction being whether IOUs are vertically integrated (table 2.4, left-hand column) or divested from generation (table 2.4, right-hand column), since I found this to be one of the key determinants of IOUs' climate policy preferences in my prior work (Basseches 2024). Once again, the seven case study states are bolded in table 2.4.

Ultimately, case selection was determined by looking at the information in tables 2.1–2.4 and selecting a sample of states that would vary, both on the outcome (RPS environmental strength) and on each of the three explanatory variables (partisanship, energy economy, electric sector structure) to enable paired comparisons guided by the aforementioned theories. Data availability was also a consideration in selecting states. Although one could legitimately argue with the case selection criteria detailed earlier, it is consistent with guidance on best practices in qualitative research methods. As Goodman (2024, 64) notes, a range of acceptable case selection methods "accommodate the oftentimes idiosyncratic nature" of "good" qualitative

Table 2.3
State-level RPS policies categorized by in-state fossil fuel production

High	Medium	Low
California	Illinois	Connecticut
Colorado	Arizona	Delaware
New Mexico	Indiana	Hawaii
North Dakota	Kansas	Iowa
Ohio	Michigan	Maine
Oklahoma	Montana	Maryland
Pennsylvania	Virginia	**Massachusetts**
Texas	Utah	Minnesota
		Missouri
		Nevada
		New Hampshire
		New Jersey
		New York
		North Carolina
		Oregon
		Rhode Island
		South Carolina
		South Dakota
		Vermont
		Washington
		Wisconsin

research. Furthermore, it is important to note that my goal in case selection was not causal analysis but rather obtaining "inferential leverage" through purposive case selection decisions (Simmons and Smith 2021).

Data Collection and Methodological Approach

Once the seven states were selected, I built in-depth, policy-focused case studies for each one that enabled process-tracing within cases and comparison across cases (Basseches et al. 2024; George and Bennett 2005). My starting point was the actual legislative/regulatory texts associated with each RPS policy in each state. With the help of a team of research assistants, I used policy reports, legislative databases such as LexisNexis State Net, and state agency websites to locate the relevant bill numbers and PUC dockets, and, ultimately, the policy texts. My team and I observed and documented each substantive change across all available versions of each policy. We did

Table 2.4
State-level RPS policies categorized by electric sector structure

Vertically Integrated IOUs	IOU Generation Divestiture
Arizona	**California**
Colorado	Connecticut
Hawaii	Delaware
Iowa	Illinois
Indiana	Ohio
Kansas	Maine
Minnesota	Maryland
Missouri	**Massachusetts**
Montana	Michigan
Nevada	New Hampshire
New Mexico	New Jersey
North Carolina	New York
North Dakota	**Pennsylvania**
Oklahoma	Rhode Island
Oregon	**Texas**
South Carolina	
South Dakota	
Utah	
Vermont	
Virginia	
Washington	
Wisconsin	

Note: Although the Pennsylvania restructuring statute did not require IOU generation divestiture, as a practical matter, all major Pennsylvania IOUs ultimately found it advantageous to divest, and did so, as state policies created favorable financial conditions for them to do so. Therefore, I placed Pennsylvania in the right-hand column.

the same for restructuring bills/statutes in states that considered/adopted them and any associated restructuring-era PUC rulemaking.

Our next step was to create a database of organized political actors that expressed a substantive preference (regarding one or more policy provisions) or a simple position (such as support/oppose) on a pending piece of legislation or a pending regulation. We accomplished this using two types of sources. The first was newspaper coverage, which we searched systematically during the period in which these policy changes were being considered. This produced a list of actors and organizations that could be followed up on, in archival research as well as through interviews. The archives themselves provided a second valuable source of actor and organization names, as well as oftentimes information about substantive policy preferences or positions. In some cases, this took the form of documentary sources (e.g., written legislative testimony), and in other cases, this took the form of audio/video footage of committee hearings, meetings, and floor debates. Table 2.5 shows the number of pages of documents reviewed in each state as part of the archival research.[16]

Once a list of actors and their representatives was compiled, the individuals were contacted for policy-focused interviews, with some "snowball sampling" (Weiss 1995) used to identify a few actors that were apparently involved behind the scenes despite no record in the archives or newspapers. Of 436 individuals reached out to across the seven states, 219 (or roughly 50 percent) agreed to be interviewed. Table 2.6 shows the distribution of interviewees, both by state and by category. Interviews were recorded and transcribed, and transcripts were coded using ATLAS.ti. Analysis of interview transcripts and archival documents followed a grounded

Table 2.5
Number of pages of documents reviewed in each case study state

California	Massachusetts	Oklahoma	Oregon	Pennsylvania	South Carolina	Texas
2,620	959	5,510	1,256	3,930	857	1,584

Table 2.6
Distribution of interviews (N=219)

State	Legislative Actors	Executive Branch Actors	Environmental/ Consumer Group Actors	Business Actors	Other Actors	Totals by State
CA	7	10	10	11	5	43
MA	14	5	13	6	0	38
OK	6	5	2	7	1	21
OR	6	5	8	7	4	30
PA	3	7	11	9	2	32
SC	7	2	6	9	2	26
TX	6	4	6	11	2	29
Totals by Category	49	38	56	60	16	**219**

theory approach (Charmaz 2006). Throughout the analysis and coding process, my research assistants and I wrote analytical memos to assist in revising preliminary conclusions as additional data was processed and analyzed. Interviewees were asked about their personal involvement throughout the policy's development, their preferences throughout the policymaking process, and their strategies for advancing those preferences. For each provision that emerged/changed/disappeared, I aimed to gather three crucial pieces of information from each respondent: *What did they want? How did they attempt to get what they wanted? And, to the degree they did not get what they wanted, whose preferences held sway instead (in their view)?*

This database—of policy texts, 16,716 pages of documents, audio/video footage, and 219 policy-focused interviews—provides a rich basis for understanding and triangulating the policy preferences and influence of a wide range of organized political actors, especially those described in chapter 1. As we will see in the coming chapters, those actors with the most nuanced and sophisticated policy preferences tended to be the IOUs, and their preferences also tended to be the most influenced by the degree to which the state had pursued restructuring of its electric utility sector.

Plan for Rest of Book

Next, we turn to the empirical analysis of the data collected. Chapter 3 explores the degree to which partisanship explains the outcomes. The data suggests that partisanship was important in setting expectations about what was or wasn't even on the table when it came to RPS strength in states like South Carolina and Oregon. However, in explaining why Texas, despite being a red state, still managed to adopt a moderately strong RPS, partisanship mattered far less than the structure of the electric sector and how it informed IOUs' policy preferences.

Chapter 4 explores the degree to which energy economy explains the outcomes. The data suggests that energy economy is helpful in explaining why a mandatory RPS was off the table in Oklahoma. However, to understand why Pennsylvania adopted a moderately strong RPS policy even though it is such a high fossil fuel–producing state, it is once again necessary to consider how the structure of the electric sector informed IOUs' policy preferences.

In chapter 5, I use evidence from multiple states to demonstrate that, overall, the single most powerful type of business actor when it came to shaping the design of the RPS policies was the IOUs. The preferences of "traditional business" actors, including fossil fuel interests, and of "green business actors," including renewable energy companies, were relatively consistent across the states and carried varying degrees of sway with policymakers, depending on the state. However, IOUs' preferences were more variable, highly dependent on the structure of the electric sector, and carried significant weight in all seven of the states. That said, the weight of IOUs' anti-RPS preferences was lessened in the states that had gone the farthest in restructuring.

Chapters 6 takes a step back from the specifics of these seven RPS cases to consider broader lessons and implications, both for the study of corporate political power and for future climate policymaking in the United States. In this chapter, I theorize the sources of IOUs' unparalleled political influence, drawing on the literature on regulatory capture. I argue that, in some ways, what makes IOUs so unusual among business actors in the climate and energy policy field is the same thing that makes them uniquely powerful.

I also consider the implications of this research for future climate policymaking efforts in the United States. The story here is both pessimistic and optimistic. On the more pessimistic side, barring a major statutory change from the federal government, the prospect of which seems particularly unlikely at the time of this writing, IOUs will likely continue to need to be "bought off" to support climate policy. Given their political power, passing climate policy in the absence of their support will be quite difficult.

On the more optimistic side, though, we are already seeing some successful state-level efforts to reduce their political power and, while this is very much approaching the problem in a piecemeal rather than holistic manner, it is still unprecedented and therefore worthy of our attention. The other optimistic takeaway is that, to the degree climate policy is about political economy and business preferences and not *just* partisanship, it suggests the possibility that progress (even if more modest) *can* be made at the state level, even with Republicans in charge of government.

3 Partisanship and Its Limits

American politics and policymaking have, over the past few decades, become increasingly characterized by hyperpartisanship, with the policy positions of the two parties on a range of issues growing increasingly farther apart (Pierson and Schickler 2024). State-level politics and policymaking are no exception, and neither is climate and renewable energy policy (Grumbach 2022). Indeed, as noted earlier, there is no stronger and more consistent predictor of state-level climate and renewable energy policy adoption in the literature than partisanship (Bromley-Trujillo et al. 2016; Coley and Hess 2012; Haseloff 2024; Trachtman 2020).

Political ideology and partisanship are two distinct concepts, with the former being about political attitudes and values and the latter being about political organization. However, political scientists have noted that in recent decades the two have become increasingly correlated, with those holding a more conservative ideology being more likely to identify as Republican and those holding a more liberal ideology being more like to identify as Democrat (Abramowitz 2022). There is some evidence that this coupling is asymmetric and more pronounced on the conservative-Republican side than on the liberal-Democrat side (Lupton et al. 2020).

The states studied in this book have in some cases flipped which party has controlled their government over the course of the 1995–2021 period. Voter ideology is likely more sticky than partisan control of state government but, nevertheless, is correlated with partisanship during the period under study. When it comes to the relationship between voter ideology and legislative behavior—which is a function not only of legislators' aim to represent the views of their constituents but also of their aim to serve party leaders' agendas and therefore gain/maintain institutional influence—Kirkland (2014) finds support for the theory that states with more ideologically heterogeneous populations will have more ideologically extreme party agendas.

This chapter uses qualitative data analysis from Oregon, South Carolina, and Texas to tease apart the specific ways in which political partisanship, and associated partisan ideologies, did and did not matter relative to the structure of the state's electric utility sector when it came to these three states' RPS policies. First, I review the background and policy history of Oregon and South Carolina. I then show that, despite both being non-restructured (vertically integrated) states with low levels of in-state fossil fuel production, Oregon's Democratic partisanship enabled a much stronger RPS policy than what was possible in South Carolina. I then review the background and policy history of Texas. In doing so, I show that despite being a "red state" with high levels of in-state fossil fuel production that limited the environmental strength of its RPS policy regime, Texas was nevertheless able to adopt a moderately strong RPS. I argue its ability to do so was largely due to its fully restructured electric utility sector, which divided the policy preferences of IOUs, and reduced the overall strength of their opposition.

Oregon Background, Policy History, and RPS Strength

Oregon, the twenty-seventh most populous US state with a population of approximately 4.2 million, is characterized by its predominantly rural geography, punctuated by a few cities such as Portland—which is by far the largest—as well as Eugene, Salem, Gresham, Hillsboro, and Bend (US Census Bureau 2021). Like many states, Oregon's rural areas vote mostly Republican and its urban areas vote mostly Democrat, but given Oregon's small overall population, the center of political gravity is Democratic, and nowadays it can safely be considered a "blue state." Despite this, given the state's political geography, its voters are ideologically heterogeneous, resulting in stronger incentives for Democratic legislators to align with party leaders' policy agendas (Kirkland 2014).

Oregon's state government was not always controlled by Democrats. While Oregon has had consecutive Democratic governors since 1987, its legislature was often controlled by the Republican Party, until the early 2000s. Beginning in 2003, the Senate moved to split party control for two years and then shortly after to Democratic Party control in 2005, where it has remained ever since. The "blue wave" hit the state House of Representatives a couple of years later, as a result of the 2006 election. Therefore, in 2007—not coincidentally, the same year the Oregon RPS was initially adopted—the House, too, came under the control of Democrats, and it has stayed that way ever since. Table 3.1 shows the partisan control of Oregon's state government from 1995–2021.

Oregon's in-state fossil fuel production is extremely low, averaging at just 0.9 trillion British thermal units (Btu) annually between 1995–2021. (By way of comparison, Texas, the highest fossil fuel–producing state in this study, averaged more than 12,699 trillion Btu annually during the same period.) Oregon produces no coal nor crude oil within its borders and only an extremely modest amount of natural gas, which accounts for all of its very limited in-state fossil fuel production. Therefore, it is classified as a "low" fossil fuel–producing state, as figure 2.2 shows. Instead, Oregon is distinctive in its reliance on legacy hydroelectric power (a low-carbon, nonintermittent electricity resource), given the state's geography, which includes the Columbia River Basin. Many of Oregon's hydroelectric dams date back to the New Deal era—long before climate change was on anybody's radar—and over the last few decades hydroelectricity has consistently accounted for more than half of the state's total electricity generation (Northwest Power and Conservation Council 2023; US Energy Information Administration 2022, 2023). As we will see, this was important for Oregon's vertically integrated IOUs' reluctant acceptance of the state's RPS.

Oregon is, however, home to a sizable coalition of "traditional business" actors (Basseches 2024), associated with industries that prior research has shown have tended to lobby *against* stringent climate and renewable energy policy, including timber, pulp and paper, and cement and metals manufacturing. At the same time, politically progressive populations in Portland and its suburbs, as well as sizable "college towns" such as Eugene and Bend, fuel a grassroots climate and environmental movement that has been a major force behind advancing state policy, including advocacy organizations such as the Oregon League of Conservation Voters (OLCV), the Oregon Environmental Council (OEC), and Environment Oregon. According to one measure of state-level environmental interest group strength—Sierra Club membership per one thousand in the state population (Newmark and Witko 2007)—Oregon is number one in the nation (Basseches 2024).

Table 3.1
Partisan control of Oregon's state government, 1995–2021

Year	Senate (Republicans/ Total Percentage)	House (Republicans/ Total Percentage)	Governor (Party)
1995	14/30 (46.7)	32/60 (53.3)	John Kitzhaber (D)
1996	19/30 (63.3)	34/60 (56.7)	John Kitzhaber (D)
1997	19/30 (63.3)	34/60 (56.7)	John Kitzhaber (D)
1998	20/30 (66.7)	31/60 (51.7)	John Kitzhaber (D)
1999	20/30 (66.7)	31/60 (51.7)	John Kitzhaber (D)
2000	17/30 (56.7)	35/60 (58.3)	John Kitzhaber (D)
2001	17/30 (56.7)	35/60 (58.3)	John Kitzhaber (D)
2002	16/30 (53.)	32/60 (53.3)	John Kitzhaber (D)
2003	15/30 (50)	35/60 (58.3)	Ted Kulongoski (D)
2004	15/30 (50)	35/60 (58.3)	Ted Kulongoski (D)
2005	12/30 (40)	33/60 (55)	Ted Kulongoski (D)
2006	12/30 (40)	33/60 (55)	Ted Kulongoski (D)
2007	11/30 (36.7)	29/60 (48.3)	Ted Kulongoski (D)
2008	10/30 (33.3)	29/60 (48.3)	Ted Kulongoski (D)
2009	12/30 (40)	24/60 (40)	Ted Kulongoski (D)
2010	12/30 (40)	24/60 (40)	Ted Kulongoski (D)
2011	14/30 (46.7)	30/60 (50)	John Kitzhaber (D)
2012	14/30 (46.7)	30/60 (50)	John Kitzhaber (D)
2013	14/30 (46.7)	26/60 (43.3)	John Kitzhaber (D)
2014	14/30 (46.7)	26/60 (43.3)	John Kitzhaber (D)
2015	12/30 (40)	25/60 (41.7)	Kate Brown (D)
2016	12/30 (40)	25/60 (41.7)	Kate Brown (D)
2017	13/30 (43.3)	25/60 (41.7)	Kate Brown (D)
2018	13/30 (43.3)	25/60 (41.7)	Kate Brown (D)
2019	12/30 (40)	22/60 (36.7)	Kate Brown (D)
2020	12/30 (40)	22/60 (36.7)	Kate Brown (D)
2021	11/30 (36.7)	23/60 (38.3)	Kate Brown (D)

Note: Middle columns of table show percentage of seats in each legislative chamber controlled by Republicans. Lighter shading indicates a simple majority for either party while darker shading indicates a two-thirds supermajority for either party. Source is Council of State Government's "Book of States."

When it comes to the electric utility sector, two major IOUs dominate Oregon's vertically integrated system: Pacific Power (PP) and Portland General Electric (PGE). PP is a multilayer subsidiary (Prechel 2020), owned by PacifiCorp, which in turn is owned by Berkshire Hathaway Energy, which in turn is owned by Berkshire Hathaway, a multinational conglomerate known for other things in addition to energy (e.g., real estate) whose CEO is Warren Buffett. On the energy side, PacifiCorp also holds Rocky Mountain Power, an IOU operating in Idaho, Utah, and Wyoming. Clearly, PacifiCorp has experience operating in states with very different partisan profiles. As one of its lobbyists told me, "The way I describe it is, it's one big company with this footprint that straddles the bluest of the blue states and the reddest of the red states across the West" (author's interview, 2/26/18). PGE, by contrast, was owned by Enron until that company went bankrupt in the early 2000s, after which TXU Energy attempted to acquire it, but that attempt was rejected by the Oregon Public Utility Commission (PUC). Instead, PGE became its own independent, publicly traded corporation. Together, these two IOUs serve more than 98 percent of Oregon's IOU customers, with the remainder served by Idaho Power (US Energy Information Administration 1995, 2001, 2008, 2021).

In general, IOUs serve about 75 percent of Oregon's residential, commercial, and industrial electricity customers. Publicly owned utilities (POUs), which include municipal utilities and public utility districts (PUDs), have served approximately 13–15 percent of these customers during the period under study. The remainder are served by rural electric cooperatives (US Energy Information Administration 1995, 2001, 2008, 2021).

When the restructuring wave hit the states in the mid-1990s, Oregon policymakers and utility industry players mostly sat and watched what was going on in other states, such as Oregon's southern neighbor, California. In 1995, the year before California adopted major restructuring legislation, the average cost of a kilowatt hour of electricity for a residential customer in Oregon was just 5.49 cents compared to 11.61 cents in California. For industrial customers—who were the real lobbying force behind the state-level restructuring movement—the gap was smaller but nonetheless significant: 7.37 cents in California compared to just 3.47 cents in Oregon. The difference between California and Oregon could likely be attributed to several factors, but chief among them would be the abundance of relatively inexpensive hydroelectricity in Oregon (and in Washington, another state that never underwent complete restructuring, where rates were even lower than in Oregon; US Energy Information Administration 1996).

Nevertheless, not wanting to be left out of the restructuring wave completely, in 1999, the Oregon legislature did enact, and Governor John Kitzhaber did sign, SB 1149, which can be considered "partial restructuring" legislation. This law enabled commercial and industrial customers to shop around for their electricity, if they could get a lower rate outside of their IOU, but critically, IOUs maintained their ownership of generation assets (in other words, they remained vertically integrated) and remained in control of retail electricity choice programs for residential customers (Oregon Senate Bill 1149, C-Engrossed 1999). Table 3.2 shows how Oregon compares with the other states in this chapter when it comes to the three major explanatory factors this book explores: partisanship, energy economy, and utility sector structure.

By 2007, Democrats were firmly in control of both chambers of the state legislature, and Democratic Governor Ted Kulongoski was eager to advance renewable energy policy with his newfound Democratic trifecta. Senate Bill 838, sponsored by Portland-based Democratic State Senator Brad Avakian, became law on June 13, 2007, and with it, Oregon joined the group of states that had adopted an RPS policy. The 2007 RPS applied only to "large utilities," defined as serving at least 3 percent of the state's total retail electricity load—a definition deliberately meant to exclude or exempt most of the state's POUs and electric cooperatives, who were a politically influential constituency opposed to the RPS due to cost concerns; however, importantly, the statute applied to both of the state's major, vertically integrated IOUs: PP and PGE (Oregon SB 838/Act 301 2007).

Table 3.2
Oregon, South Carolina, and Texas: Explanatory factors

	Partisanship	Energy Economy	Utility Sector Structure
Oregon	"Blue"	Low fossil fuel production	Vertically integrated
South Carolina	"Red"	Low fossil fuel production	Vertically integrated
Texas	"Red"	High fossil fuel production	Hyper-restructured

For the utilities included, the 2007 RPS Act stipulated that at least 5 percent of their portfolio must be generated from qualified renewable electricity sources in years 2011–2014. That number ratcheted up to at least 15 percent for the years 2015–2019, then to at least 20 percent for the years 2020–2024, and finally to at least 25 percent for the years 2025 and beyond. As an Oregon IOU lobbyist told me, "The 25 by 2025 was obviously a bumper sticker" (author's interview, 2/26/18). He continued:

> In fact, I remember testifying. . . . There was a point where based on our [holding company's] Iowa experience, we thought that getting above 15% was going to be very tough, [but] part of it is . . . we knew it was important to establish the RPS to the proponents, and so getting the framework in place is more important than what happens ten or fifteen years from now. There will be other legislatures, there are other mechanisms in here to call a timeout if things go out of whack, and so we weren't overly concerned about it. (author's interview, 2/26/18)

The "other mechanisms" included §12 of the 2007 Act, which stipulated a "cost cap"; that is, if the incremental cost of compliance with the RPS were at any point in the future to equal more than 4 percent of the utility's total "revenue requirement" (all operating expenses plus the company's rate base multiplied by its PUC-sanctioned rate of return for the IOUs' shareholders), the utility would be excused from compliance (Oregon SB 838/Act 301 2007).

In terms of what counted as "qualified renewable electricity sources," the 2007 Act stipulated eligible technologies to include wind, solar, wave, tidal, and geothermal. With respect to more environmentally controversial technologies, such as hydroelectricity and biomass, they were eligible but with limitations. In the case of hydroelectricity—which readers will recall has long been an abundant electricity source in the Pacific Northwest—a compromise was struck that electricity generated from hydroelectric facilities that became operational or experienced significant upgrades (including the environmental designation of "low impact") on or after January 1, 1995, would be eligible. This was critical to gaining vertically integrated IOUs' eventual support. Regarding biomass, it would be largely eligible except for municipal solid waste and wood treated with certain chemicals. Additionally, the facility generating the electricity—regardless of the technology—would need to be geographically located within the jurisdiction of the Western Electricity Coordinating Council (WECC[1]; Oregon SB 838/Act 301 2007).

Unbundled renewable energy credits (RECs) were allowed for compliance purposes, but their use was limited to 20 percent of a utility's total compliance obligations. Importantly to the IOUs, the 2007 Act permitted the unlimited banking of RECs, affording regulated utilities maximal flexibility in how to use RECs accumulated over time. The law required the Oregon PUC to establish a penalty in the form of alternative compliance payments (ACP) and provided for PUC discretion in creating additional penalties beyond that (Oregon SB 838/Act 301 2007).

In terms of its environmental strength, Oregon's RPS was aggressive for its time. While some twenty-one states beat Oregon chronologically in terms of their date of initial RPS adoption, Oregon's nominal targets, especially for the years 2015 and beyond, were at least as aggressive as those of other climate policy leading states like Massachusetts, Connecticut, and Rhode Island (Barbose 2023). However, Oregon's law—like those of all the climate-policy-leading-states in this study—included certain provisions added solely on account of IOUs' preferences, which allowed any costs/risks of compliance to be shifted onto their customers (Basseches 2024).

During the 2009 legislative session, with Kulongoski still the Democratic governor and with Democrats having significantly grown their majority in the state House of Representatives as a result of the 2008 election (likely riding the coattails of Barack Obama, who bested John McCain in Oregon by over sixteen points), Oregon enhanced its RPS. In particular, the legislature passed and the governor signed HB 3039, which established a residential rooftop solar pilot program within the state's RPS regime. To nullify IOUs' opposition to the legislation, language was inserted clarifying that IOUs—not the participating households—would own the associated RECs and that they could be used to comply with the IOU's overall obligations under the RPS. Furthermore, a provision, in §3, stated that any associated compliance costs would be recoverable in rates and eligible for the "automatic adjustment clause" mechanism, which IOUs had secured previously (Oregon HB 3039/Act 748 2009; Basseches 2024).

During the following legislative session, in 2010, legislation was adopted that slightly weakened the RPS from an environmental standpoint by expanding the eligibility of hydroelectricity and biomass resources. Specifically, HB 3649, signed into law in 2010, relaxed certain requirements for qualifying hydroelectric facilities (Oregon HB 3649/Act 71 2010), while HB 3674, adopted in the same year, allowed electricity generated from municipal solid waste to be eligible for up to 11 average megawatts of RPS compliance beginning in the year 2026, even if the facility was operational prior to 1995 (Oregon HB 3674/Act 17 2010). These were provisions that made Oregon's vertically integrated IOUs more comfortable in their ability to comply. HB 3691, also adopted in 2010, fortified cost-shifting provisions for IOUs (Oregon HB 3691/Act 79 2010).

The next significant enhancement of Oregon's RPS came in 2016. SB 1547, which passed that year in a Democratic trifecta with margins in the legislature similar to 2009, required the elimination of all in-state coal-fired electricity generation by January 1, 2030. As a practical matter, this statute was targeted squarely at the one remaining coal-fired power plant in the state, the PGE-owned Boardman plant. To nullify PGE's potential opposition, §1 stipulates: "The Public Utility Commission shall adjust any schedule of depreciation approved by the commission for an electric company's coal-fired resource . . . [such that] the adjusted depreciation schedule . . . require[s] the coal-fired resource to be fully depreciated on or before December 31, 2030."

The same law increased and extended the RPS targets and timelines for the IOUs such that they would now need 27 percent eligible renewable resources by 2025 (instead of 25 percent) and targets would be established for 2026–2040, ultimately ratcheting up to 50 percent by 2040. The same law further relaxed resource eligibility requirements for biomass facilities, including municipal solid waste, and repealed the solar carve-out. On the other hand, it placed limitations on REC banking for IOUs for the first time in the statute's history (Oregon SB 1547/Act 28 2016). In 2021, under Democratic Governor Kate Brown and a Democratic legislature with even more comfortable margins, the state adopted HB 2021, which created a brand-new clean electricity standard extending to 2040 but did not alter the existing RPS statute (Oregon HB 2021/Act 508 2021). Even though, as this section shows, RPS policy design is complex and ever-evolving, table 3.3 shows how Oregon's RPS compares to the other states' RPSs analyzed in this chapter in terms of a few of the key elements of policy design.

South Carolina Background, Policy History, and RPS Strength

South Carolina has a population comparable to Oregon's. Slightly larger, it is the twenty-third most populous US state, with approximately 5.1 million

Table 3.3
Oregon, South Carolina, and Texas: RPS outcomes (selected policy design features)

	Mandatory or Voluntary?	Nominal RPS Target Percentage (Barbose 2023)	Regulated Parties	Controversial Resource Eligibility
Oregon	Mandatory	50–74	"Large" investor-owned utilities	Existing hydroelectricity and biomass (but with limitations)
South Carolina	Voluntary	N/A	N/A	"Low-impact hydro" and biomass
Texas	Mandatory	<25	Retail electric providers (REPs)	"New" hydro and biomass only

residents. But unlike Oregon, it has fewer and less populous politically liberal cities. Its largest city, Charleston, has a population of just 150,000 (3 percent of the state population) compared to Portland, whose population of 650,000 constitutes some 15 percent of the state's total (US Census Bureau 2021a). Consequently, South Carolina is very different from Oregon politically; among other things, South Carolina is much more ideologically homogenous than Oregon (Kirkland 2014).

Indeed, South Carolina's House of Representatives has been controlled by Republicans ever since 1996, and its Senate has been the same ever since 2002. Ever since Democrat Jim Hodges departed the governor's mansion and Mark Sanford took over, in 2003, the state's governors have been consecutively Republican up until the time of this writing. Currently, Republicans are just shy of a supermajority in the state legislature. Table 3.4 shows the partisan control of South Carolina's state government from 1995–2021.

In terms of its in-state energy resources, South Carolina is, like Oregon, a "low" fossil fuel–producing state, as shown in figure 2.2. In fact, South Carolina produces virtually no fossil fuels within its borders, and is therefore even lower than Oregon, which produces a tiny amount of natural gas (US Energy Information Administration 2023). Unlike Oregon, hydroelectricity represents only a small fraction of South Carolina's in-state electricity generation, and instead, a sizable portion of South Carolina's electricity is generated by nuclear power plants (US Energy Information Administration 2022).

Table 3.4
Partisan control of South Carolina's state government, 1995–2021

Year	Senate (Republicans/ Total Percentage)	House (Republicans/ Total Percentage)	Governor (Party)
1995	16/46 (34.8)	50/124 (40.3)	David Beasley (R)
1996	19/46 (41.3)	65/124 (52.4)	David Beasley (R)
1997	19/46 (41.3)	65/124 (52.4)	David Beasley (R)
1998	21/46 (45.7)	71/124 (57.3)	David Beasley (R)
1999	21/46 (45.7)	71/124 (57.3)	Jim Hodges (D)
2000	22/46 (47.8)	67/124 (54)	Jim Hodges (D)
2001	22/46 (47.8)	67/124 (54)	Jim Hodges (D)
2002	25/46 (54.3)	71/124 (57.3)	Jim Hodges (D)
2003	24/46 (52.2)	73/124 (58.9)	Mark Sanford (R)
2004	25/46 (54.3)	73/124 (58.9)	Mark Sanford (R)
2005	26/46 (56.5)	74/124 (59.7)	Mark Sanford (R)
2006	26/46 (56.5)	74/124 (59.7)	Mark Sanford (R)
2007	26/46 (56.5)	73/124 (58.9)	Mark Sanford (R)
2008	27/46 (58.7)	73/124 (58.9)	Mark Sanford (R)
2009	27/46 (58.7)	71/124 (57.3)	Mark Sanford (R)
2010	27/46 (58.7)	73/124 (58.9)	Mark Sanford (R)
2011	26/46 (56.5)	76/124 (61.3)	Nikki Haley (R)
2012	27/46 (58.7)	76/124 (61.3)	Nikki Haley (R)
2013	28/46 (60.9)	77/124 (62.1)	Nikki Haley (R)
2014	28/46 (60.9)	78/124 (62.9)	Nikki Haley (R)
2015	28/46 (60.9)	78/124 (62.9)	Nikki Haley (R)
2016	28/46 (60.9)	78/124 (62.9)	Nikki Haley (R)
2017	28/46 (60.9)	80/124 (64.5)	Henry McMaster (R)
2018	28/46 (60.9)	79/124 (63.7)	Henry McMaster (R)
2019	26/46 (56.5)	79/124 (63.7)	Henry McMaster (R)
2020	27/46 (58.7)	80/124 (64.5)	Henry McMaster (R)
2021	30/46 (65.2)	81/124 (65.3)	Henry McMaster (R)

Note: Middle columns of table show percentage of seats in each legislative chamber controlled by Republicans. Lighter shading indicates a simple majority for either party while darker shading indicates a two-thirds supermajority for either party. Source is Council of State Government's "Book of States."

South Carolina's interest group politics were also similar to Oregon's in terms of the overrepresentation of "traditional business" actors relative to "green business" ones (Basseches 2024). In particular, the steel industry and the automotive industry were prominent "traditional business" actors in South Carolina, with an interest in the status quo when it came to the state's energy policy. Additionally "the Chamber of Commerce, Americans for Prosperity [and others] . . . parachuted in" (author's interview, 5/18/22) to engage in the state's energy policy discussions (author's interviews, multiple). Still, given the enormous solar energy potential of the US Southeast, a number of entrepreneurial "green business" actors pushed successfully for pro-solar policies in the mid-2010s. However, a mandatory RPS was never on the table in South Carolina. The voluntary policy that emerged, according to one clean energy advocate, "was completely imperfect and it didn't move the needle in any grand way, but it ultimately kind of set the stage for a lot of the progress that happened since then" (author's interview, 5/19/22).

In terms of environmental advocacy organizations, it was surprising to learn while doing fieldwork on the ground in the state capital that several politically savvy environmental groups have been active there and have even found pathways to real policy influence. These include the Coastal Conservation League (CCL), the Conservation Voters of South Carolina (CVSC), and the Southern Environmental Law Center (SELC). For instance, when I interviewed a staffer for the state legislature's Labor, Commerce, and Industry committee, which had jurisdiction over the state's voluntary RPS legislation enacted in 2014, I was told (about the influence of these environmental groups): "It was The League [CCL] and "Joe"[2] [that initiated the legislation], I mean that's how I got to know him. . . . Joe was at the [legislative Chairman's office] door, you know, just making noise. . . . The League [CCL] . . . Southern Environmental Law Center, certainly, probably drafted more [of the bill] than anybody" (author's interview, 6/7/22).

These environmental groups and their lobbyists were acutely aware of the political environment in which they were operating and deployed a very different strategy than their counterparts in "blue states" did. As one South Carolina environmental advocate explained: "You will never see the words "climate change" in any of the messaging we did for the Energy Freedom Act[3]. And I still have advocates that will complain to me about that and I'm like, "Dude, we got it done. Like, do you want me to do something, or do you just want me to be symbolic?" . . . It was about jobs and economy and energy independence and, like, sticking it to the utility" (author's interview, 5/23/22).

Speaking of utilities, two major IOUs, Duke Energy and Dominion Energy,[4] dominate in South Carolina, serving 96–100 percent of the state's IOU customers over the period of 1995–2021.[5] Both Duke and Dominion are large energy parent companies operating multiple subsidiaries in multiple states. Importantly, the majority of the territorial footprint of both companies is in the US Southeast, one of two regions (the other being the West with the exception of California) where IOUs are vertically integrated and the electricity sector was not significantly restructured (see chapter 1). In South Carolina, an average of approximately 55 percent of electricity customers over the period 1995–2021 were served by IOUs, with the remainder served by POUs and rural electric cooperatives, especially the latter (Ford and Bailey 2022; US Energy Information Administration 1995, 2001, 2008, 2021). This is a noteworthy distinction from Oregon, where POUs and cooperatives constituted a smaller share of total customers. Table 3.2 shows how South Carolina compares with the other states in this chapter when it comes to the three major explanatory factors this book explores: partisanship, energy economy, and utility sector structure.

During the restructuring era, South Carolina—like the larger Southeast region—did not act. Electricity rates for all customer classes were relatively low compared to most of the states that *did* restructure, though not quite as low as they were in the Pacific Northwest (US Energy Information Administration 1996). Vertically integrated IOUs were extremely politically and economically concentrated and entrenched and were unwilling to entertain any efforts to strike a deal with pro-competition forces on restructuring (Harrison and Welton 2021). As a legislative staffer in South Carolina put it: "There were staffers, like myself, who were young and dumb and would say things about deregulation or competition. . . . But you know, that was shut down so quickly because it was a nasty word. . . . Deregulation was a dirty word. . . . [Efforts to restructure] got shot down so quickly by Duke, Dominion. . . . You know, yeah, those were just dirty words" (author's interview, 6/7/22).

An interview with an IOU lobbyist in South Carolina confirmed the IOUs' strategy there to do everything possible to avoid any move toward restructuring or deregulation whatsoever: "I was told to watch out for the possibility of decoupling.[6] That was not something we were interested in. . . . Decoupling . . . makes the utility strategically more vulnerable to deregulation . . . it sort of sets you up [for that]" (author's interview, 5/22/22).

South Carolina's IOUs, especially the predecessor to Dominion, South Carolina Electric and Gas, were weakened politically by a scandal involving VC Summer, a $9 billion nuclear power plant in Fairfield County whose construction was never completed due to cost overruns, despite South Carolina ratepayers having already paid for it thanks to a 2007 state law that "the utilities wrote" called the Base Load Review Act (author's interviews, multiple; Holdman 2023).[7] As one South Carolina state legislator put it: "You know, the one thing about crises or catastrophes like that, Josh, is that it provides a window of opportunity to change things from the way you've always done them" (author's interview, 6/1/22).

This *relative* (and perhaps fleeting) weakening of South Carolina IOUs politically, and the "window of opportunity" it provided, yielded two concrete legislative results. The first was the passage of the Energy Freedom Act, which, despite having been written in such a way that "the lobbyists for those two utilities [Duke and Dominion] got most of what they wanted" (author's interview, 6/7/22), was ultimately a win for climate and renewable energy advocates relative to the status quo because it expanded incentives such as net metering for rooftop solar. The second result, more relevant to the present study, is that it enabled the passage of HB 4940, which created a legislative committee to *study* restructuring in South Carolina and ultimately produce a report making policy recommendations for the state to pursue (South Carolina HB 4940/Act 187 2020).

The report, which the legislature retained the consulting services of The Brattle Group to prepare, was published on April 27, 2023. Its recommendations were clear: "At present, South Carolina's electricity sector is structured under a vertically integrated utility model. . . . We find that South Carolina ratepayers stand to gain substantial benefits from a measured introduction of enhanced regional coordination and market reforms" (Tsoukalis et al. 2023, 1–2). Yet at the time of this writing, the legislature has taken no such recommended action, a testament to the enduring political power of the IOUs (Basseches 2024; Culhane et al. 2021; Harrison and Welton 2021; Stokes 2020).

Despite the lack of a general restructuring policy in South Carolina, large industrial customers have still managed to reduce their electricity cost burdens by striking bilateral deals with the IOUs, which have reduced their electricity bills relative to those of residential and small, commercial customers. This has created an unlikely alliance in the state between IOUs and the types of business actors that in other work I have referred to as "traditional business" (Basseches 2024): energy intensive and/or trade-exposed industrial electricity customers. While in other states, IOUs and these "traditional business" actors were often battling one another when it came to renewable energy policy design and especially when it came to rate cases before state PUCs (Basseches 2024), in South Carolina, they were more often on the same team in front of policymakers, bolstering the political and economic power of one another while maintaining the status quo to the detriment of smaller electricity customers. When asked about this dynamic, a South Carolina energy lobbyist explained: "Well, part of it is that Duke and SCE&G [Dominion] were big, influential members of the organized Manufacturers' Alliance" (author's interview, 5/19/22). A former legislative staffer added: "They [large industrial customers] have relationships with their utilities that serve them. And they're very sophisticated deals that they have. . . . And they don't want to upset the apple cart [by opposing the IOUs in policymaking discussions]" (author's interview, 5/19/22). It is in this context—a "ruby red" state in terms of political partisanship (unlike Oregon), with an energy economy not reliant on fossil fuel production (similar to Oregon), and politically and economically influential, vertically integrated IOUs (similar to Oregon)—that we must seek to understand South Carolina's purely voluntary RPS policy.

In 2014, SB 1189 was signed into law by Republican Governor Nikki Haley (South Carolina SB 1189/Act 236 2014). It was a heavily negotiated legislative product, which followed the convening of a wide range of energy policy stakeholders through a multiyear process that was called the "Energy Advisory Council (EAC)" (author's interviews, multiple; South Carolina Legislature's Website). In addition to several other solar energy-related provisions, this law—which started out as a bill authored by Republican State Senator Chauncey "Greg" Gregory—created the "South Carolina Distributed Energy Resource Program," also known as the state's voluntary RPS. Under this voluntary program, an IOU such as Duke or Dominion had the *option* of applying to the South Carolina Public Service Commission (SCPSC)[8] with a plan that "shall, at a minimum, result in development by

2021 of renewable energy facilities located in South Carolina in an aggregated amount of installed nameplate generation capacity equal to at least two percent of the previous five-year average of the electrical utility's South Carolina retail peak demand" (South Carolina SB 1189/Act 236 2014, §2). Critically, for this legislation to be adopted despite the effective veto power of South Carolina's politically entrenched IOUs (author's interviews, multiple; Harrison and Welton 2021), any utility opting in to such a program "shall be permitted to recover its costs related to the approved program" (South Carolina SB 1189/Act 236 2014, §2).

Furthermore, the 2014 South Carolina law included other provisions affecting the distributive politics of the state's voluntary RPS policy in IOUs' favor. These included the stipulation that "the effect of a decision to amend or terminate an approved DOER program shall be prospective only, and costs incurred prior to that decision shall be recoverable" (South Carolina SB 1189/Act 236 2014, §2). In addition, "An electrical utility may invest in distributed energy resources or programs outside of an approved distributed energy resource program under this chapter. . . . *The fact that such resources are not part of an approved distributed energy resource program shall create no negative inference concerning their recoverability under other ratemaking provisions* [emphasis added]" (South Carolina SB 1189/Act 236 2014, §2).

South Carolina's voluntary RPS program has not been amended since, though the Energy Freedom Act of 2019, which passed in the aftermath of the VC Summer debacle, was considered a win for clean energy and climate advocates, both "green business" actors (Basseches 2024), such as the Southern Alliance for Clean Energy (SACE) and E4 Carolinas, and environmental groups, such as the aforementioned CCL, CVSC, and SELC. Although IOUs Duke and Dominion continued to exercise effective veto power over the language of this legislation (author's interviews, multiple), green business and environmental wins included expansion of the state's net metering program (especially important to the solar industry) and energy efficiency measures, including enhanced requirements for IOUs' integrated resource plans (IRPs). The law also included consumer protections aimed at preventing a repeat of the VC Summer debacle (South Carolina HB 3659/Act 62 2019). Table 3.3 shows how South Carolina compares to the other two states analyzed in this chapter when it comes to certain key elements of RPS policy design.

Explaining Oregon Versus South Carolina: The Role of Partisanship and Ideology

Given the similarities between Oregon and South Carolina in terms of their energy economies not including significant fossil fuel production and their IOUs being vertically integrated—owning generation, transmission, and distribution assets and not being part of an RTO—how can we best understand why Oregon adopted a relatively strong RPS policy while South Carolina adopted an entirely voluntary and environmentally weak RPS policy? In this section, I will argue that the differences in partisan politics (and associated differences in political ideology) between the two states best explain their divergent outcomes. This is not surprising considering the existing literature (e.g., Carley and Miller 2012; Coley and Hess 2012; Egan and Mullin 2023; Haseloff 2024), though my analysis offers a new look at qualitative mechanisms undergirding policy design differences.

In particular, I identify three partisanship-/ideology-related factors that created a very different political environment in these two states. The first is the timing of the Oregon RPS's initial adoption in 2007, precisely the session when Democrats took control of both legislative chambers and had a supportive Democratic governor in Ted Kulongoski, as well as the policy's strengthening as the size of the Democratic legislative majority grew in the years that followed. The second is an ideological distinction, which closely reflects partisan differences (Lupton, Smallpage, and Enders 2020), between a "mandate-driven approach" to renewable energy policy (as Oregon's stringent, mandatory RPS is) and a more "market-oriented approach" (as was preferred in South Carolina). Finally, the third is the fact that, in Oregon, grassroots-powered environmental advocacy organizations, closely aligned with the majority (Democratic) party, succeeded in applying pressure to adopt a stringent RPS in a way that was not possible in South Carolina.

The Partisan Origins and Gradual Strengthening of the Oregon RPS

As mentioned earlier, while Oregon's RPS is among the environmentally strong ones analyzed in this book, the state was not an early adopter. Indeed, twenty-one states adopted an RPS before Oregon did. The state senator who authored the 2007 legislation was very aware of this fact. As he told me: "I was concerned at the time that Oregon was I think the 22nd state to pass renewable portfolio standards. Something like that. We were not the first.

And we were a state that I knew should be a leader, so I wanted to pass an aggressive and effective standard" (author's interview, 2/8/18). It was not a coincidence that 2007 was the year that the initial RPS was adopted. This coincided with the first legislative session in which the Democratic Party controlled both chambers of the legislature plus the governor's office; a Democratic trifecta in state government.

Another Oregon state legislator, who played a major role in shepherding SB 838 through the House of Representatives, highlighted the importance of Democrats' newfound legislative majority in 2007:

> **Respondent:** "Overall, there was a lot of pent-up energy—no pun intended—around environmental issues. I joined the legislature in 2001 and served for six years in the minority party. Pretty much I got one bill through. Actually I think I got two. One was a task force and the other, I think it was a study. So we got a lot of issues that certainly myself and Sen. Avakian worked on that we were just waiting for the time to be right [to pass]."
>
> **Interviewer:** "Meaning, for the Democrats to have the majority?"
>
> **Respondent:** "And chairing the committees, yeah."[9] (author's interview, 2/13/18)

A different Oregon state legislator, also an environmental champion, concurred:

> That session was sort of unusually successful when it came to environmental legislation. . . . There was a lot of stuff going on. . . . One was that in the 2006 election Democrats took the majority in the House, which they hadn't held for, I don't know, eight or ten years? Maybe longer. As a result, there had been a number of policies that were just—had been hung up by the Republican majority. (author's interview, 2/9/18)

Consistent with political science research emphasizing the importance of majority party leadership in controlling the legislative agenda (e.g., Anzia and Jackman 2013; Curry 2015; Rosenthal 2009), this same legislator talked about the pressure his colleagues felt at the time to stay aligned with the Democratic Speaker, who was a strong supporter of this legislation, especially given the narrow margins of his party's legislative majority. He explained:

> The 31/29 majority created a dynamic where Democrats kind of had to stick together, and there was a high price to pay for any Democrat who did not stick with the Caucus. . . . If you were the Democrat who said, "Oh, I'm not sure about that priority" . . . if you were the Democrat who threatened to vote with the Republicans, you knew it was a slippery slope. . . . As a whole, the Democratic caucus had

> a lot of things it wanted to get done. . . . Jeff Merkley [the Speaker of the House at the time] . . . was a great team leader. I mean, he helped establish this dynamic that I'm describing, and he created a real sense of camaraderie and . . . a sense of unity. (author's interview, 2/9/18)

In that 2007 session, the Democrats' majority was slim, and party leaders' goal was simply to get an RPS on the books. But as Democrats' legislative majorities grew larger in subsequent sessions, they seized opportunities to strengthen the RPS. This was particularly true of the 2009 and 2016 legislative RPS updates described in a previous section of this chapter. "Now Democrats had a much larger majority 36/24," a state legislator said when asked about the update in the 2009 session (author's interview, 2/9/18). Consistent support from Democratic governors—referred to by Rosenthal (2009, 264) as "chief legislators"—was also helpful throughout, as Democratic Governor Ted Kulongoski "really believed we needed to move to a renewable energy regime" (author's interview, 2/14/18), and Democratic Governor Kate Brown championed the major update in 2016, which eliminated coal-fired electricity generation from Oregon's portfolio (Anderson 2016).

In South Carolina, where Republicans have had a firm grip on both legislative chambers and the executive branch since 2002, there was no political opportunity to accomplish anything close to what we saw in the case of Oregon. When asked why anything more stringent than a voluntary RPS was off the table in South Carolina, a renewable energy lobbyist told me: "I don't think Republicans in the legislature wanted a state-wide energy mandate" (author's interview, 5/18/22). A different one told me, specifically referencing the timing of SB 1189, which presumably would have been the best chance for something mandatory: "The Tea Party happened . . . and it completely changed the political landscape around energy issues in the state. . . . And so, even at the local level in Charleston, we were working with the City on a green plan that just got shelved" (author's interview, 5/19/22).

Free-Market Ideology: No Mandates in South Carolina

Not only can we see a distinction between the two states in terms of how the partisan composition of state government affected the prospects for renewable energy policy, but we can also see how a more cultural or ideological difference in values—closely correlated with partisan differences (Lupton et al. 2020)—made its way into the policymaking discussions in the two states. A significant reason a mandatory RPS was off the table in

South Carolina had to do with the state's dominant political ideology, promoting freedom and markets.

Both major clean energy laws that passed in South Carolina—the bill that included the voluntary RPS (SB 1189/Act 236) and the Energy Freedom Act of 2019—deliberately emphasized freedom, consumer choice, and free market principles. They certainly did not emphasize climate change, and extra care had to be taken to ensure that the policies would under no circumstances be framed as "mandates." A renewable energy lobbyist explained the strategy:

> We knew we wanted a Republican to be the lead on [the legislation], and push it as a consumer choice, free market—like, this isn't telling anybody to do anything. All this is doing is providing access and it's allowing access for people that also otherwise aren't going to have that kind of cash [to own solar panels]. (author's interview, 5/19/22)

A different renewable energy lobbyist explained the deliberate choice of wording about what to call the 2019 Energy Freedom Act: "I mean, the 2019 Energy *Freedom* Act, I mean, that word was intentionally included in the title of that Act because it plays well politically. . . . Essentially, what you're talking about is personal liberty [emphasis added]" (author's interview, 5/20/22).

This renewable energy lobbyist went on to explain how, while sensitive to political messaging considerations, the act was also "truth-in-advertising," referring specifically to the provisions regarding the solar leasing program: "[It's about] letting people who *want* to put solar panels on their roof do it, even if they have to lease the panels, right? I mean, why are we not allowing that? So yeah, it's more personal freedom and individual liberties, yeah [emphasis in original]" (author's interview, 5/20/22).

A utility lobbyist suggested that a mandatory RPS would be politically untenable in South Carolina simply because of the *mandatory* part. South Carolinians, according to this lobbyist, do not like policies that originated elsewhere, such as a mandatory RPS, being imposed on them: "RPS was a dirty word . . . because it came from away, it came from other states" (author's interview, 5/22/22). While it is true that the vertically integrated IOUs in South Carolina were strongly opposed to a mandatory RPS policy for economic, self-interested reasons, one interviewee suggested that the utilities' preferences would not have mattered anyway, because South

Carolinians, and their elected representatives, would have been unwilling to consider one: "We weren't going to get an RPS. I mean, even if the utilities want an RPS, we wouldn't be able to get one in South Carolina [because of the political ideology here]" (author's interview, 5/19/22).

By contrast, in Oregon, the mentality was the opposite. *Because* other states that were similarly politically progressive had adopted strong RPS mandates already, it was Oregon's time to do the same. "RPSs were all the rage in the mid-2000s," one Salem lobbyist told me (author's interview, 2/26/18). A former Kulongoski administration official noted the importance of the fact that both California and Washington had already adopted mandatory RPS policies: "California was putting a lot of pressure [for us to adopt one too]" (author's interview, 2/26/18).

Environmental Organizations and Partisan Alignments

Finally, as Trachtman and Meckling (2022) show, there is strong alignment between Democrat-led states and climate advocacy. In my study, it was notable not only how active and influential environmental interest groups were in Oregon relative to other states but also the degree to which they were aligned with elected leaders in the Democratic Party. This is not to say that environmental advocates in South Carolina were ineffective; to the contrary, as noted previously, it was surprising how politically nimble South Carolina's environmental groups were when it came to adapting to their political environment and, frankly, enjoying more influence than I was expecting to find when I entered the field. Still, unlike in Oregon, where there was a tight alliance between such groups and Democratic politicians (indeed, organizations like the Oregon League of Conservation Voters [OLCV] were influential in getting many of them elected), in South Carolina, environmental groups found that Democratic politicians, besides being in the minority, were unreliable as allies to a renewable energy agenda. Therefore, these groups often tried working with Republicans instead, but their ability to advance stringent policy was limited by factors already discussed.

In Oregon, it was helpful that both the Senate and House champions of the 2007 RPS bill came directly out of the environmental advocacy community themselves, illustrating the importance of network ties: "So you had the two relevant committees both chaired by people who were directly from the environmental community. And so, Brad [State Senator Brad Avakian] was the co-founder of OLCV's Washington County chapter; Jackie [State Representative Jackie Dingfelder] was a former board member of OLCV. And so this was a driving mission of theirs" (author's interview, 2/18/18).

While this was a happy coincidence that certainly greased the wheels for SB 838, the reality is that the environmental groups in Oregon were extremely unified in their agenda and had extremely close ties with members of the new Democratic majority more generally (author's interviews, multiple). That session, they had formed a coalition, together with consumer organizations such as the Citizens Utility Board (CUB), to advocate for the RPS: "The Fair and Clean Energy Coalition at that time had over 100 organizations . . . the Oregon League of Conservation Voters, the Oregon Environmental Council, CUB . . . Northwest Energy Coalition . . . Renewable Northwest Project. Low-income groups" (author's interview, 2/9/18).

A leading advocate I interviewed described the RPS as "the top thing" in his "eight-point agenda" (author's interview, 2/9/18) for that 2007 legislative session. Their unified advocacy was helpful to legislators. A state senator, speaking about these environmental and consumer groups, explained: "For many, many years they all had their own agendas and were working their own bills. And they came together as a coordinated team that session. And so, you had the strongest team of advocates that I had ever seen all on the same page on the same bills and with tight agreement that they weren't going to counter each other" (author's interview, 2/8/18).

By contrast, in South Carolina, the environmental community found that, on renewable energy policy in particular, Democrats were not reliable allies. Instead, they found themselves working with Republicans selectively but having to greatly limit their policy ambitions and having to tread carefully so as not to be perceived as "climate activists." One professional environmental advocate described the strategy: "I have legislator friends who will not vote on anything until they have run it by their political consultant. We have deep relationships with some red [Republican] political consultants, and so we were also coordinating with them" (author's interview, 5/23/23).

It was surprising what little success these advocates had in working with Democratic legislators who, despite being in the minority party in South

Carolina, are the usual allies of these types of interest groups across the fifty states (Trachtman and Meckling 2022). In one case, the failure of a renewable energy bill in South Carolina was attributed to the groups having selected the "wrong sponsor" to shepherd the legislation, who was the wrong choice *because* he was a Democrat (author's interview, 5/23/23). Another more structural complication that impeded what we might have expected to be a more productive alliance between environmental groups and Democratic politicians in South Carolina was the racial and geography-based politics of the state and how they were exploited by utility interests. According to one environmental advocate:

> A lot of people come to South Carolina from anywhere else, and they just assume that we [environmentalists] have the Democratic vote. We do *not* have the Democratic vote on environmental issues because there are two things that dominate partisan politics in South Carolina. There's a racial division. There's only one Black Republican, there are very few White Dems. . . . There's also a big divide between urban and rural. . . . And the utilities intentionally contract with all the Black lobbyists. . . . They have turned a lot of the Black Democrats against solar because they think it's hurting low-wealth or low-income residents [emphasis in original]. (author's interview, 5/23/23)

And so, partisan politics and associated differences in political ideology certainly played the most significant role in explaining why Oregon and South Carolina ended up with such divergent RPS regimes, despite both being states with limited in-state fossil fuel production and politically entrenched, vertically integrated IOUs. I have illustrated three mechanisms, observable in qualitative data, through which partisan politics operated in these two cases. None of this is to say that partisanship was the only factor at play in these states nor that it is impossible to adopt a mandatory RPS in a "red state." To the contrary, as we will soon see, Texas adopted a moderately strong mandatory RPS policy despite such partisan conditions. And certainly, when it comes to *public* attitudes on climate and renewable energy, there is spatial and within-party variation in policy preferences (Ansolabehere and Konisky 2016; Mildenberger et al. 2017). Nevertheless, partisanship and associated ideological differences played the greatest role when examining the cases of Oregon and South Carolina, side by side.

However, as we will soon see, partisanship is hardly the only, or even the most important, factor at play when it comes to state-level RPS policy design. Examining the case of Texas shows that partisan obstacles can be overcome by favorable interest group politics. In particular, a restructured electric utility sector in which IOUs were no longer vertically integrated monopolies created such conducive political conditions.

Texas Background, Policy History, and RPS Strength

Texas is the second most populous US state and is extremely demographically diverse, despite being a major stronghold of the Republican Party. Its largest cities, including Houston, San Antonio, Dallas, and Austin, vote Democratic. However, these four cities, which constitute approximately 20 percent of the state's total population, are politically underrepresented, given partisan districting and the political strength of Texas's rural areas, which are vast, as the nation's second-largest state in square mileage (US Census Bureau 2021b). Table 3.5 shows the growth over time in Republican Party strength in the state legislature and the consistent control of its executive branch by the Republican Party, from 1995–2021 (and continuing at the time of this writing). Like South Carolina, prior to about 2002, control of the state legislature was more contested between the two major parties.

At an average of 12,699 trillion Btus annually during the period 1995–2021, Texas produces more fossil fuel than any other US state, and it's not even close, with Wyoming coming in a distant second, at an annual average of 8,300 trillion Btus during the same period. Geologically blessed with the Permian Basin, the plurality of Texas's fossil fuel production is natural gas, followed closely by oil, followed less closely—but still in relative abundance—by coal (US Energy Information Administration 2023). Texas is the quintessential "high" fossil fuel–producing state and is indicated as such in figure 2.2.

Correspondingly, the lion's share of Texas's electricity mix came from gas and coal during the period under study. For instance, in 1995, these two technologies alone generated 86 percent of Texas's total electricity. By 2021, thanks to an influx of wind energy, which came online in the intervening years in part thanks to the state's moderately strong mandatory RPS policy (Wiser and Langniss 2001), the gas and coal quotient was down to 67 percent, which is still sizable compared to other states during the same period. Hydroelectricity and nuclear have played a much smaller role in Texas's electricity mix compared to in Oregon and South Carolina,

Table 3.5
Partisan control of Texas state government, 1995–2021

Year	Senate (Republicans/ Total Percentage)	House (Republicans/ Total Percentage)	Governor (Party)
1995	13/31 (41.9)	58/150 (38.7)	George W. Bush (R)
1996	14/31 (45.2)	63/150 (42)	George W. Bush (R)
1997	14/31 (45.2)	63/150 (42)	George W. Bush (R)
1998	17/31 (54.8)	68/150 (45.3)	George W. Bush (R)
1999	17/31 (54.8)	68/150 (45.3)	George W. Bush (R)
2000	16/31 (51.6)	71/150 (47.3)	George W. Bush (R)
2001	16/31 (51.6)	71/150 (47.3)	Rick Perry (R)
2002	15/31 (48.4)	71/150 (47.3)	Rick Perry (R)
2003	19/31 (61.3)	88/150 (58.7)	Rick Perry (R)
2004	19/31 (61.3)	88/150 (58.7)	Rick Perry (R)
2005	19/31 (61.3)	87/150 (58)	Rick Perry (R)
2006	19/31 (61.3)	87/150 (58)	Rick Perry (R)
2007	20/31 (64.5)	81/150 (54)	Rick Perry (R)
2008	20/31 (64.5)	79/150 (52.7)	Rick Perry (R)
2009	19/31 (61.3)	76/150 (50.7)	Rick Perry (R)
2010	19/31 (61.3)	77/150 (51.3)	Rick Perry (R)
2011	19/31 (61.3)	101/150 (67.3)	Rick Perry (R)
2012	19/31 (61.3)	101/150 (67.3)	Rick Perry (R)
2013	19/31 (61.3)	94/150 (62.7)	Rick Perry (R)
2014	18/31 (58.1)	95/150 (63.3)	Rick Perry (R)
2015	20/31 (64.5)	98/150 (65.3)	Greg Abbott (R)
2016	20/31 (64.5)	98/150 (65.3)	Greg Abbott (R)
2017	20/31 (64.5)	95/150 (63.3)	Greg Abbott (R)
2018	20/31 (64.5)	93/150 (62)	Greg Abbott (R)
2019	19/31 (61.3)	83/150 (55.3)	Greg Abbott (R)
2020	19/31 (61.3)	83/150 (55.3)	Greg Abbott (R)
2021	18/31 (58.1)	82/150 (54.7)	Greg Abbott (R)

Note: Middle columns of table show percentage of seats in each legislative chamber controlled by Republicans. Lighter shading indicates a simple majority for either party while darker shading indicates a two-thirds supermajority for either party. Source is Council of State Government's "Book of States."

respectively, though Texas receives significantly more of its electricity from nuclear sources than from hydroelectricity (US Energy Information Administration 2022).

Given this energy economy, it is unsurprising that the "traditional business" actors (Basseches 2024) in Texas were numerous and politically influential. My interview and archival data are filled with examples of this. Trade associations such as the Texas Association of Manufacturers (TAM), the Texas Association of Business (TAB), and the Texas Industrial Energy Consumers (TIEC) all represented oil and gas interests, among others, and, of course, many oil and gas companies also lobbied individually against pro-renewable energy policy. As one lobbyist explained: "You had a legislator and a chair of the committee . . . whose staff was very personally close with the lobbyists of [one of the aforementioned associations]" (author's interview, 3/2/22).

Think tanks such as the Texas Public Policy Foundation[10] were busy protecting such interests in terms of public messaging and media strategy, while lobbyists associated with the aforementioned groups, as well as in-house lobbyists at individual companies—ExxonMobil and Occidental Petroleum (rebranded as "Oxy") were mentioned specifically (author's interview, 3/11/22)—were busy working the "inside game" with policymakers. All of these interests were forceful in fighting against an environmentally strong RPS in the state. As one lobbyist told me: "Texas Public Policy Foundation is a great example where, you know, they come out and they start blaming wind [for everything]. It's like, 'I have a cold and I sneezed. It's wind's fault'" (author's interview, 3/18/22).

That being said, Texas's "green business" actors (Basseches 2024) *did* exist in the late 1990s and early 2000s when the Texas RPS was created and then expanded, despite being nascent and overpowered politically, relative to the oil and gas industry (Stokes 2020). In particular, two coalitions of such green business actors, the Texas Renewable Energy Industries Association (TREIA) and The Wind Coalition, were instrumental in teaming up with environmental groups and organizing politically to establish an RPS in the state. They even managed to hire some high-profile lobbyists, with ties to George W. Bush and Karl Rove, among others (author's interviews, 3/10/22, 3/11/22, 3/22/22).

Although they faced an uphill battle trying to convince Texas policymakers to adopt strong, pro-renewable policies (Stokes 2020), they benefitted from certain advantages, as well. For one, there was enormous untapped potential that onshore wind in particular enjoyed in large swaths of rural Texas, including West Texas and the Panhandle. These areas, whose voters were mostly Republican, benefitted from disproportionate representation in the state legislature and also from local economic development opportunities associated with an increased presence of the wind energy in their communities (author's interviews, multiple). A Republican state senator, who was a key player in the passage of the initial RPS and authored the subsequent bill to later enhance it, described himself as one such "rural legislator" (author's interview, 3/22/22). He explained:

> One of my reasonings for establishing the [RPS] amendment—it was an economic decision for the State of Texas, because we were going into this with gas prices at an all-time high. We had about 40%—maybe more, we got about 50% of our generation was natural gas, and now gas prices were at an all-time high. I was looking for economic stability in the pricing structure . . . and I thought that building wind in Texas was both a natural thing to do, because we've got a lot of wind, and I thought it would work. But it was an economic decision. . . . A lot of the area where the wind generation is, there's also oil and gas there. And you'd have a farmer with a cotton field, with pumpjacks pumping oil, wind generators that were right beside it, sticking up into the air, and you'd see a tractor plowing or planting cotton. They understand very clearly that when you have three revenue streams, it helps a lot. . . . I was from that area, and I recognized the potential. (author's interview, 3/22/22)

Although environmental and consumer interest groups often struggled to influence policy in a place like Texas relative to a place like Oregon, or certainly relative to other more politically progressive states like California, New York, or Massachusetts, these groups were nevertheless present in Texas in the late 1990s as the RPS was getting started and still enjoyed a seat at the table in policymaking discussions. Renewable energy lobbyists as well as public interest group advocates each described their alliance in those early days as critical and mutually beneficial. A renewable energy lobbyist described a prominent public interest group advocate at the time as the individual "most responsible" for getting the RPS included in Texas's 1999 restructuring legislation (author's interview, 3/11/22) while that same public interest group advocate described that same renewable energy lobbyist as "our secret weapon" (author's interview, 3/11/22).

National environmental and consumer public interest groups found Austin (the state capital) to be a strategically beneficial place to invest resources, given Texas's size and its likelihood of setting an example for other states or even the federal government. The Environmental Defense Fund (EDF), which established "one of their major regional offices" (author's interview, 3/2/22) in Austin, was one such group. One of its leaders, who was active over the years on restructuring and the RPS, explained:

> At the time, EDF was founding some state offices. . . . Some environmentalists in Texas that were friends of mine thought EDF would be a good fit for Texas, . . . So, we approached . . . EDF about the possibility and basically, they said "If you can raise a fair amount of money upfront from Texas sources to fund most of the first three years of operation, we'll do it." (author's interview, 3/10/22)

They did, and along with the consumer group Public Citizen, they became a significant force in the establishment of Texas's RPS. As a corporate lobbyist representing industrial customers at the time told me, "Public Citizen and EDF are the two big ones, so they had a seat at the table" (author's interview, 2/25/22).

Texas's major IOUs are far more numerous than in Oregon and South Carolina. There are multiple reasons for this. One is Texas's sheer size. Another, however, is that the state's restructuring law required the separation of what were, at the time, single vertically integrated IOUs, into three separate levels of separate companies, associated with each of the three levels of the electricity system described in chapter 1. While these companies were allowed to be affiliated with one another through multilayer subsidiary structures (within the bounds of certain restrictions intended to guard against market power), they needed to be separate operating companies, and they tended to have separate names (Texas SB 7 1999).

Under Texas's restructuring law, IOUs would need to either be: 1) exclusively in the business of generating wholesale electricity to be sold competitively to electric retailers ("generating utilities"); 2) exclusively in the business of owning the transmission and distribution systems ("transmission and distribution utilities"); or 3) exclusively in the business of selling electricity competitively to end users ("retail electric providers"). Importantly, these new rules applied only to the region of Texas that is the Electric Reliability Council of Texas (ERCOT), as opposed to the comparatively small regions of the state that are part of the other RTOs, the Midcontinent

Independent System Operator (MISO), and the Southwest Power Pool (SPP; Texas SB 7 1999, §39.051).

IOUs belonging to the first category, the generating utilities, would no longer be regulated by the state's PUC, since they would not operate as monopolies but rather as competitive wholesale generators. IOUs belonging to the second category, the transmission and distribution utilities, would remain monopolies and be traditionally regulated through cost-of-service regulation, like in Oregon and South Carolina. And finally, IOUs belonging to the third category, REPs, would *not* be monopolies either and would remain somewhat overseen by the PUC, but would also be allowed to set their own rates competitively, so as to compete for customers. (Texas SB 7 1999).

Today, myriad competitive generators—including some owned by parent companies also owning regulated utilities in other states, such as Florida Power and Light (FP&L) and Avangrid—operate in Texas. The largest transmission and distribution utilities operating in the state include subsidiaries of American Electric Power, CenterPoint Energy, Entergy Texas, El Paso Electric, Oncor, and Texas–New Mexico Power. And the largest REPs include Calpine, Constellation, Direct Energy, NRG, and Reliant (Public Utility Commission of Texas 2024). Table 3.2 shows how Texas compares with the other states in this chapter when it comes to the three major explanatory factors this book explores: partisanship, energy economy, and utility sector structure.

Over the years under study, IOUs—whether those outside of ERCOT serving retail customers or REPs competing for customers within ERCOT—have served approximately 68–70 percent of Texas's electricity customers. POUs, primarily municipally owned utilities (including large ones like Austin Energy and San Antonio's CPS Energy), have served approximately 15 percent. The remaining approximately 15 percent have been served by rural electric cooperatives (US Energy Information Administration 1995, 2001, 2008, 2021).

The political origins of restructuring in Texas were somewhat different from those of other restructured states discussed in subsequent chapters, where high electricity prices were the driving force. Interestingly, while electricity rates in 1995 were higher for residential and commercial customers in Texas than they were in Oregon (which opted not to restructure), they were on par with those in South Carolina (which also opted not to restructure). And in 1995, industrial customers in Texas were already enjoying lower rates than in any other US state that opted for restructuring (US Energy Information Administration 1996).

So why did Texas policymakers opt for restructuring? The short answer is politics. Two political forces were at play: corporations and ideology. On the corporation side, you had Enron, headquartered in Houston, Texas, and developing a business plan of retail power marketing that required restructuring, state by state. Few markets were as big as Enron's home state of Texas. As one Texas state legislator at the time recalled: "Ken Lay [Enron's CEO] was still alive and extremely popular. Huge charitable donor. . . . He'd fund the whole [state legislative intern] program. . . . And then people would be handed out—whether you were Democrat or Republican, you know, you got somebody on your staff from this program" (author's interview, 2/28/22).

An Enron lobbyist at the time recalled: "I remember it well . . . I was on vacation in June of 1998 when the phone call came and said, "Oh my god, we've got a Texas legislative session and our Texas guy just left. Would you come back and do Texas electric for us?" . . . There is going to be a [restructuring] bill and it is going to get passed and we [Enron's leadership] need to make sure that it is the best possible bill that we can have" (author's interview, 3/17/22).

There were also an unusually large number of politically powerful industrial customers in Texas—including oil, gas, and chemical companies—that were not content with some of the lowest electricity rates in the country at the time; they wanted lower. And they had the political muscle to get lower. As one electricity regulator at the time put it, they were "jumping on the Enron train" (author's interview, 3/17/22).

On the ideological side, you had Governor George W. Bush, and PUC Chairman Pat Wood, whom Bush appointed, who believed very strongly in the promise of free markets to deliver optimal results for consumers. Wood was the primary architect and was widely described by those I interviewed—of varying political stripes—as a "genius" and a "genuinely good person" (author's interviews, multiple). For example, a state senator who worked with him on the restructuring legislation described him as "instrumental," "the expert," and "just intricate to the whole thing" (author's interview, 3/12/22). It was Governor Bush who empowered Wood, and Wood recalls having the following conversation with him:

> He [Bush] said, you know, "There's a lot of big stuff going on at the PUC . . . There's an open spot there" . . . And so I asked him, I said, "So what are your objectives and why do you think—why do you want this competitive market?"' And he said, "Pat, the utilities," and he meant the electric and the phone, "They care more about what *we* think . . . than what their customers think. And that's wrong, and we're going to change that across the board. So that's what your job is. Do you want it? [emphasis in original]." (author's interview, 3/18/22).

More generally, as a member of Wood's staff put it, he and Bush both subscribed to "the whole Reagan magic, and view of competition and consumer choice as God-given rights" (author's interview, 3/17/22).

However, while it was a Republican-held ideology that, in part, provided the impetus for Texas's restructuring law, the legislation itself was bipartisan, or even nonpartisan. SB 373, which became law in 1995 and laid the groundwork for the major restructuring bill, SB 7, in 1999, passed a Senate and House that at the time were still controlled by Democrats, and the bill was signed into law by a Republican (George W. Bush). By the time the 1999 legislative session arrived, Republicans had taken control of the Senate, and SB 7, the major restructuring bill, was championed by Republican State Senator David Sibley and Democratic State Representative Steve Wolens, in the Senate and House, respectively. It, too, was signed into law by Republican Governor George W. Bush (Texas Legislature Online 2024). As Commissioner Wood put it, "It just isn't a partisan thing, with regard to utility regulation. Or at least it wasn't [back] then" (author's interview, 3/18/22).

SB 373 laid the groundwork for comprehensive restructuring in Texas by authorizing transmission-level, wholesale restructuring and by making organizational changes at the PUC that would facilitate competition (Texas SB 373 1995). In 1997, there was an attempt at complete restructuring, including retail competition, but that failed due to the political power of POUs and rural cooperatives, which, along with the major vertically integrated IOUs at the time, largely opposed it (author's interviews, multiple). In 1999, when SB 7 was introduced and ultimately became law, POUs' and rural cooperatives' opposition was nullified by giving them the ability to "opt in" to competition without *requiring* them to do so—an option that few, if any, of these utilities have taken at the time of this writing (author's interviews 3/8/22, 3/21/22; Texas SB 7 1999). Meanwhile, IOUs ultimately stood down in 1999, after negotiating several generous provisions with legislators.

The most important of these was a "nonbypassable competitive transition charge," which would allow IOUs to recoup all so-called stranded costs from ratepayers after an initial "rate freeze" period (Texas SB 7 1999, §39.201(g)). Stranded costs represent the remaining value of the generation assets that IOUs were required to divest from under restructuring. Specifically, §39.252 of subchapter F of the new chapter 39, which SB 7 added to the Utility Code, establishes a "right to recover" such costs. While the utility executives of the early 1900s would have never imagined this when they were negotiating the original "regulatory compact," they would likely have approved, given the irresistible push toward competition in Texas and elsewhere (Hirsh 1999).

Critically, the exact value of these stranded costs would be adjudicated at the PUC, a forum in which IOUs have traditionally enjoyed even greater influence than in the state legislature, in part because they control most of the data that PUC regulators rely upon to make such decisions (Kovvali and Macey 2023). The statute provided utilities great financial flexibility in accounting for the value of such assets. For instance, §39.256 gave them the "option to re-direct depreciation expenses related to transmission and distribution assets to its net generation plant assets" while subchapter G provided for "securitization," offering the rationale that "this type of debt will lower the carrying costs of the assets relative to the costs that would be incurred using conventional utility financing methods" (Texas SB 7 1999, §39.301).

Vertically integrated IOUs in Texas did not embrace restructuring initially, but they ultimately accepted it, and some saw it as an opportunity to expand upon and innovate upon their traditional way of doing business. As one lobbyist told me: "They actually believed that they might do better under a competitive market than a regulated market" (author's interview, 3/10/22). The political power of "traditional business" actors (Basseches 2024) in Texas, largely stemming from Enron and the fossil fuel industry's unusually large presence in the state, forced IOUs to the negotiating table in a way not seen elsewhere.

That being said, once at the table, IOUs did quite well for themselves, as IOUs tend to do (Basseches 2024). A PUC regulator described some of the associated negotiations like this:

> The price of that stock went down 15% that day. All the Texas utilities' [stock price] went down that day. That night, they were in Bush's office. . . . The CEOs of the three major utilities were in the Governor's office saying . . . "We've got to

> get [our stranded costs recovered]"—And Bush was very good . . . he didn't interfere in any of this stuff. But he just, he called us down and he said, "I'm not telling you about how to do your job" he smiled and he said, "I think it's time to start cutting a deal. . . . Get to work." (author's interview, 3/18/22)

The deal that resulted was a "grand set of bargains" (author's interview, 3/17/22), according to one of the negotiators. And according to this same negotiator, "In the case of the utilities in '99, they got a whopping amount of stranded cost [recovered from ratepayers], plus securitization" (author's interview, 3/17/22).

IOUs were not the only ones that saw an opportunity in SB 7, the restructuring law. Renewable energy advocates did too and brilliantly engaged in what Hurlbut (2008, 132) calls "strategic linkage," linking an RPS to the broader restructuring effort, to get the RPS included within the statute despite the Republican-controlled state senate and the Republican governor.[11] In fact, Bush, Sibley, Wolens, and Republican Senator Troy Fraser (who was on a committee with jurisdiction over SB 7 and later authored a bill strengthening the RPS in 2005) were all open-minded about including a mandatory RPS in SB 7, despite the fact that three of the four were Republicans. As one of them (a Republican) told me: "I'm glad we did [include the RPS in SB 7]. . . . I thought it was a good thing in the long run, you know, to have a diversified portfolio. And I think any diversified portfolio ought to have, I think, wind, coal, gas, *and renewables*. I mean, all of the . . . well, mostly wind I guess, so I thought that was fair, I thought it was a good policy [emphasis added]" (author's interview, 3/12/22).

A prominent public interest group advocate explained the tactic he used to keep the pressure on these policymakers to include the RPS in the negotiated statute:

> I told them early on, I'm like, "No matter what conditions, I'm not going to be able to support it [SB 7]." But others [other environmental and consumer groups] did. . . . And we helped orchestrate that . . . because we knew that it had to be done right. And having myself and some others out, you know, shooting arrows at the bill until the last minute, continuing to get it better and better, allowed Wolens the room to move [to include an RPS]. (author's interview, 2/28/22)

Even a lobbyist representing industrial electricity customers—"traditional business" actors (Basseches 2024) that opposed the RPS due to cost concerns but very much supported SB 7 overall, given their hopes of lowering electricity costs—gave the environmental advocates credit for their negotiating tactics: "They knew that we needed them to get on board and they felt like they could do something big here if they could, you know, work that process. And I think they were right" (author's interview, 2/25/22).

This sort of insider lobbying tells only part of the story (albeit the part most proximate to the drafting of the actual legislative language). External factors mattered, too. One such factor, also discussed by Hurlbut (2008), was "deliberative polling" that was conducted in the late 1990s showing widespread support for renewable energy among Texan residential electricity customers, even if they had to pay more for it. Urban areas like Dallas were experiencing heavy air pollution at the time, which may have been a factor affecting these results. According to multiple interviewees, the results of the polling caught IOUs off guard, weakening their ability to oppose the RPS outright, and also weakening consumer advocates' argument against restructuring on the whole (with an RPS included), given clear consumer demand (author's interviews, multiple).

The 1999 RPS was modest in terms of its targets and timelines. At the time, roughly 880 MW of renewable generation capacity existed in the state, and §39.904 of the new Utility Code chapter created by SB 7 required 2,000 MW of *additional* renewable capacity by January 1, 2009. The statute also specified additional interim goals for 2003, 2005, and 2007. The details of resource and facility eligibility requirements—and frankly, a lot of the details more generally—were deliberately delegated to the PUC, which was instructed to promulgate rules to effectuate the statute "not later than January 1, 2000." The regulated parties included IOUs (specifically, REPs in the new multilevel structure), POUs, and cooperatives. The decision to leave most of the other details of policy design to the nonelected (gubernatorially appointed, in this case) PUC commissioners was a deliberate one, and helped further insulate the policy from partisan politics (author's interviews, multiple).

The PUC took this mandate and ran with it, with its leadership acting as what Rabe (2004) accurately describes as "policy entrepreneurs." In Substantive Rule 25.173, adopted December 20, 1999, the PUC issued a ninety-five-page order specifying these important details. *Annual* (as opposed to the legislature's biannual) interim targets were added, ratcheting up to the ultimate target of 2,000 MW added by 2009. Resource eligibility was on par with most other states' RPSs (Fischlein and Smith 2013), allowing biomass (including landfill gas) but excluding existing fossil power plants "re-powered" to use a renewable fuel. Importantly, existing hydroelectricity was excluded (recall the slippery slope we saw with this in Oregon), as were *all* facilities that were preexisting at the time. Indeed, one of the PUC staffers I interviewed noted that she "threatened to quit" when her boss, one of the commissioners, wanted to "give credit to existing hydro" (author's interview, 3/17/22). She did not quit, and existing hydroelectricity was excluded (Public Utility Commission of Texas 1999), preserving some of the policy's environmental strength.

The PUC also developed a REC accounting system that allowed limited flexibility to REPs and other parties subject to RPS regulation. This included banking of up to three years but not forward-borrowing as has been seen in other states (which weakens the environmental strength of the RPS). When it came to penalties and ACPs, another important feature that can be used to evaluate RPS strength (Basseches 2024; Fischlein and Smith 2013), the PUC established the rule that "the penalty shall be the lesser of $50 per MWh or, upon suitable evidence of market value by the competitive retailer [REP], 200% of the average market value of credits for that compliance period" (Public Utility Commission of Texas 1999). Despite language like this offering some degree of relaxed compliance flexibility for IOUs (specifically, REPs, in Texas's multileveled system), when compared to what other states had in place at that time, Texas's initial RPS was still among the more stringent in the country for its time—keeping in mind that climate policy leading states like California did not even *adopt* an RPS until 2002 or later (Basseches 2024; Carley et al. 2018; Fischlein and Smith 2013). Table 3.3 shows how Texas compares to the other two states analyzed in this chapter when it comes to certain key elements of RPS policy design.

The 2005 legislative session resulted in an enhancement of the RPS; it was strengthened from an environmental standpoint. This happened despite the fact that, by then, a Republican trifecta had assumed power in Austin (see table 3.5). Indeed, it was a *Republican* state senator—self-described as having "the most conservative voting record in the Senate" (author's interview, 3/22/22)—that authored SB 20, which strengthened the RPS. He told me: "I was, you know, considered at that time what they referred to as hardcore. I was a very, very conservative Republican" (author's interview, 3/22/22).

Specifically, SB 20 extended the RPS from 2009 to 2015, requiring an additional 5,000 MW by that later date. Crucially, SB 20 contained what in other states is referred to as a "solar carve-out" (see chapter 1); in Texas, it was simply called a "non-wind" portion: "Of the renewable energy technology generating capacity installed to meet the [RPS] after September 1, 2005, the Commission shall establish a target of having at least 500 MW of capacity from a renewable energy technology other than a source using wind energy" (Texas SB 20 2005, §3).

However, while SB 20 strengthened the RPS overall, despite the Republican trifecta, it was not without its concessions from the perspective of renewable energy advocates, with two in particular. The first illustrates the political strength of both "traditional business" actors (Basseches 2024) and the IOUs (specifically, the REPs) whereas the second benefited only the IOUs (specifically, the transmission and distribution utilities).

The first was described by a renewable energy lobbyist as a "poison pill" (author's interview, 3/11/22), which the bill's proponents needed to "swallow," if the bill was going to pass at all. It has also been written about by Hurlbut (2008). Specifically, this provision—adding to the Code §39.904(m)—reduced REPs' REC requirements "by the amount of green power purchased by customers voluntarily" (Hurlbut 2008, 137; Texas SB 20 2005). This was a win for both the REPs, which now had lesser compliance obligations, and for industrial electricity customers, who were lobbying hard against *any* renewable energy mandates, fearful that they could raise electricity costs when IOUs inevitably passed them through to ratepayers (author's interviews, multiple). This provision, however, was repealed in 2007 and therefore had little to no real-world effect on renewable energy development in the state (Hurlbut 2008). The same bill that repealed it also, in §2, authorized the PUC to establish a separate alternative compliance payment (ACP) for the "non-wind" portion of the RPS and limited that ACP to no more than $20 per MWh (TX HB 1090 2007).

The second concession in SB 20 had less to do with the environmental strength of the RPS and more to do with its distributive quality (Basseches 2024). Specifically, this concession represented a shifting of costs/risks associated with the renewable energy transition away from IOUs and onto ratepayers. This took the form of two separate provisions with similar distributive effects. One provision, §1(d), required the PUC to allow transmission and distribution utilities to pass all costs associated with renewable energy transmission onto ratepayers "regardless of the extent of the utility's actual use of the facilities." Another, akin to the "automatic adjustment clause" that I documented in Oregon (see Basseches 2024), in §2, allowed these same utilities to receive automatic approval of certain transmission projects and their associated costs, borne by ratepayers (TX HB 1090 2007).

And so, while Texas's RPS policy is far from perfect from an environmental standpoint (among other things, it has not been significantly updated since 2007, despite its initial targets having long been surpassed), it is certainly much stronger than a voluntary RPS of the sort we saw in South Carolina. After all, it was mandatory, it had nominal targets and timelines that were relatively aggressive (especially for their time), and it even had what was effectively a solar carve-out (in the form of a "non-wind" portion). It also had relatively stringent resource and facility eligibility requirements. As one renewable energy advocate reminded me, crediting the RPS, "In 2001, Texas installed more wind power than the United States had in any previous two-year period. . . . That was the biggest year nationally for wind growth, and yet Texas doubled what they did in the rest of the country" (author's interview, 3/11/22).

How Texas Got It Done: The Role of a Restructured Utility Sector

As the preceding section makes clear, the secret to Texas's success was not Democratic politicians. While the restructuring law that was adopted with the RPS included in 1999 was adopted by a Democrat-controlled House of Representatives, it also made it through a Republican-controlled Senate and was signed into law by a Republican governor, George W. Bush, who would go on to lead the national Republican Party as president of the United States from 2001–2009. Furthermore, the RPS's most significant enhancement became law in 2005, in an environment that was a Republican trifecta within the state's government.

In this section, I will argue that, in addition to the "strategic linkage" of the RPS with restructuring (Hurlbut 2008), another important and understudied factor that enabled Texas's relative success was the new interest group politics that Texas forged by restructuring its IOUs—the single most influential type of political actor in state-level renewable electricity policymaking (Basseches 2024)—into three separate types of corporations: 1) generating utilities, 2) transmission and distribution utilities, and 3) REPs. The result was to divide the IOUs—which, in vertically integrated states, formed a unified, entrenched, and all-powerful lobbying bloc—in terms of their renewable energy policy preferences, thus weakening their collective opposition to the RPS outright in Texas. Political scientists often refer to this as "policy feedback effects" (Moynihan and Soss 2014). In other words, restructuring in Texas created a different interest group landscape than in both Oregon and South Carolina. In Texas, unlike in the other two, IOUs' policy preferences were more divided, and thus their collective power to block the RPS diluted, effectively split into thirds.

On the one hand, competitive generating utilities varied in terms of their attitudes toward the RPS. While some did not embrace it, others saw an opportunity to diversify their portfolios, building renewables at a subsidized cost (exactly what the RPS was intended to do). At the time, while fossil fuels, particularly natural gas, were still considered cheaper (a reality that has been turned on its head by the time of this writing, due to a confluence of factors including state and federal policies), natural gas prices were rapidly rising prior to the fracking boom of the mid-2000s. Some fossil-fueled generators saw renewables as a way to hedge against this risk. As one lobbyist told me, "Energy Future Holdings [a generating utility] . . . was very much on board and supportive [of the RPS], like not just sort of "Okay, we're not going to oppose it" but very supportive . . . They saw the opportunity to build new [renewable] generation from it" (author's interview, 3/2/22). As another example, AES Corporation, a competitive generating utility, testified in support of Senate Bill 541 in 2009, which would have created a solar RPS for Texas (Texas State Legislative Archives 2009).

Transmission and distribution utilities, for their part, were generally agnostic toward the RPS. Texas law (and the "regulatory compact," in effect in all fifty states) allows them to recover all of their costs from ratepayers, plus the regulated rate of return they are allowed for their shareholders under cost-of-service regulation (see chapter 1). As one lobbyist explained, "This didn't really affect them materially, so they weren't terribly involved [in lobbying on it]" (author's interview, 3/2/22).

This can be seen in the positions taken by TXU Energy, which, postrestructuring, became primarily a transmission and distribution utility company in Texas (with affiliated companies in the generation business). TXU Energy did not oppose an ACP as part of the Texas RPS and made this position known in deliberations at the PUC. This same company also opposed a provision under consideration by the PUC that would have allowed generating utilities to offset RPS compliance with preexisting renewable resources, obviously discouraging the creation of *new* renewable facilities (Public Utility Commission of Texas 1999). And one of their lobbyists testified *in favor* of the 2005 RPS enhancement before a Texas House committee, saying of SB 533,[12] "We think the bill is a very balanced approach" (Texas State Legislative Archives 2005).

It was truly only the REPs that were more or less uniformly opposed to the RPS. Recall it is *these* actors that are the regulated parties under the RPS statute, meaning *they* are the ones responsible for demonstrating compliance to the PUC by retiring the requisite number of RECs. For them alone, the RPS had the potential to raise the costs of doing business in a competitive retail environment, which is precisely why they effectively lobbied along with "traditional business" actors for the "poison bill" in the 2005 law discussed previously. A renewable energy lobbyist recalls one major REP in particular "showed up and testified in opposition" to the RPS legislation, and others undoubtedly worked behind the scenes (author's interview, 3/2/22). A REP lobbyist explained: "When you increase an RPS, it adds cost to the retailer [REP], which we couldn't pass along [to the customer] according to the price-to-beat formula [one of the provisions of SB 7]" (author's interview, 3/14/22). For this reason, the lobbyist explained, "We weren't crazy about it" (author's interview, 3/14/22).

We can also see REP opposition manifested in Enron's opposition to an ACP at the PUC (recall that Enron was eager to get into the competitive power marketing business at the time of restructuring; Public Utility Commission of Texas 1999). And we can see it again in the testimony that Reliant Energy, a major REP postrestructuring, supplied to a state senate committee in 2009, opposing a solar RPS and citing cost concerns (Texas State Legislative Archives 2009).

We know from past research that IOUs are, on the whole, the most influential type of private actor in state-level renewable electricity policymaking, and that they often but not always oppose such policies (Basseches 2024; Culhane et al. 2021; Stokes 2020). This was certainly the case in South Carolina, where they helped to keep a mandatory RPS completely off the table, and in Oregon, where their opposition was overcome *only* by policymakers having made numerous cost-shifting concessions (and having included generous provisions regarding the eligibility of existing hydroelectricity generation), in an environment of strong Democratic partisan control aligned with strong environmental groups and virtually no in-state fossil fuel production. The threat of a ballot initiative as was seen in Washington State was another element that brought Oregon IOUs to the table. Even so, PGE, the Oregon IOU that owned one existing in-state coal plant in 2007, was more reluctant in accepting the negotiated settlement, although it eventually came around to do so (author's interviews, multiple). However, in contrast to both South Carolina and Oregon, in Texas, only about a third of the IOU industry was actually opposed to the RPS, with the remainder being neutral to supportive.[13]

This made the internal politics within their statewide trade association, Association of Electric Companies of Texas (AECT), extremely complex and often resulted in the group not taking a position on legislation, instead relying on individual companies to lobby for their own self-interest (author's interviews, multiple). It diluted their ability to stop or weaken the RPS. As one lobbyist explained: "When you're talking about the utility before [restructuring], it was vertically integrated. Now you're talking about three separate companies, and they often don't agree with one another" (author's interview, 2/25/22).

Furthermore, there is some limited evidence in my data that the type of IOU most opposed to the RPS, the REPs, were politically weakened by no longer being regulated utility monopolies with a guaranteed rate of return, like their transmission and distribution utility counterparts, and like vertically integrated utilities in nonrestructured states. As one lobbyist told me:

> One of my favorite stories, I was going either to an ALEC [American Legislative Exchange Council] or NCSL [National Conference of State Legislatures] meeting . . . and I'm getting on the plane. . . . The Reliant [a vertically integrated IOU before restructuring and a REP after] woman [lobbyist] was getting on the plane and she said, "Yeah, I'm going to Dallas to pick up so and so, a Dallas legislator, and fly with that legislator out to the conference." . . . And she was assigned a legislator, and they had an army of people who each had their own legislator. . . . But then I did it for a couple more years and they continued to do that until the reality of retail competition [restructuring] set in and they couldn't afford to have that many people anymore. (author's interview, 3/17/22)

What this effectively meant is that the segment of the IOU industry most likely to oppose the mandatory RPS in Texas was weaker politically than its former self, prior to restructuring.

It is important not to paint an overly rosy picture of climate and renewable energy policy in Texas. As Stokes (2020) has shown, Texas is far from an ideal environment for such policies, and its hospitability when it comes to such policies has certainly deteriorated over time, as Republicans have grown their legislative majorities and the Republican Party itself has become more extreme, a trend that is observable in other states and nationally as well (Grumbach 2022; Pierson and Schickler 2024). Stokes correctly points out that Texas failed to adopt solar energy policy in the 2009 legislative session (despite at least two bills having been put forward to do so). And predictably, as noted earlier, REPs opposed such legislation, as did of course Texas's sizable and extremely powerful oil and gas industry (author's interviews, multiple).

However, in this chapter, I have presented evidence that shows a new, underappreciated, and critical factor in the political economy of RPS policy design. While in other contexts the blame for renewable energy policy shortcomings might be laid at the feet of partisanship (e.g., Vasseur 2014) or utility lobbying (e.g., Stokes 2020), the Texas case shows how much the *structure* of the utility sector matters as well. Specifically, in Texas, a combination of the less partisan nature of electric utility regulation, especially in the earlier days, the "strategic linkage" (Hurlbut 2008) of electric sector restructuring and renewable energy policy, and the divided preferences of IOUs, allowed for a moderately strong mandatory RPS policy that helped make Texas the US state with more wind-powered electricity generation than any other (Wiser and Langniss 2001). In the next chapter, we will explore further the role of in-state fossil fuel production when it comes to the interest group politics of RPS policy design. We will find it to be important, just as we found partisanship, in explaining some of the variation, but not all.

4 Energy Economy and Its Limits

In addition to partisanship, energy economy is another factor that is thought to explain variation in state-level RPS policy outcomes (Bromley-Trujillo et al. 2016; Haseloff 2024; Huang et al. 2007; Matisoff 2008; Thombs and Jorgenson 2020). Bromley-Trujillo and coauthors find that a higher share of gross state product derived from mining is negatively associated with climate policy adoption (Bromley-Trujillo et al. 2016). Thombs and Jorgenson (2020) find that states with an RPS tend to have less fossil fuel production per capita than states without one, and that a liberal government is a more important determinant of RPS adoption in high fossil fuel–producing states. And Haseloff (2024) finds that states that are more economically dependent on oil and gas production are less likely to have a mandatory RPS.

A qualitative study like this one is best equipped to determine—albeit for a smaller number of cases—not only *whether* in-state fossil fuel production limited RPS policy strength and/or affected RPS policy design but also *how,* or through what mechanisms, fossil fuel industry influence occurred. This study also allows us to tease apart the *relative* influence of in-state energy economy compared to other factors such as partisanship or electric utility sector structure. Finally, it allows us to see in which ways these factors may have interacted with one another to explain policy outcomes.

In this chapter, I argue that high levels of in-state fossil fuel production best explain why Oklahoma adopted a purely voluntary RPS policy as well as the limited environmental strength of Pennsylvania's mandatory RPS policy. At the same time, to explain why Pennsylvania was able to overcome its high-levels of in-state fossil fuel production to adopt a mandatory policy, while Oklahoma was not, I show that the differing structure of the two states' electric utility sectors was what mattered the most.

Despite Pennsylvania being the fourth largest fossil fuel–producing state in the country between 1995–2021, and the third largest coal-producing state after West Virginia and Wyoming (US Energy Information Administration 2023), its restructured electric utility sector is what enabled it to nevertheless adopt a moderately strong RPS policy.

First, I provide background on the Oklahoma case, tracing its policy history from the restructuring era to Oklahoma's adoption of its voluntary RPS policy in 2010 (which was never updated since, and allowed to expire). Next, I provide background on the Pennsylvania case, tracing its policy history from the restructuring era to the most recent update of its RPS policy in 2007 (including unsuccessful attempts to strengthen it in the years since). I then draw on my qualitative database to analyze the politics that led to these outcomes. I find that energy economy was the single most important factor in *limiting* the strength of the Oklahoma RPS and that the restructuring of Pennsylvania's electric utility sector was the single most significant factor in *enabling* that state's moderately strong mandatory RPS, which, despite its shortcomings and the failed attempts to strengthen it beyond the mid-2000s, was nonetheless an impressive and puzzling achievement when viewed in light of the extant literature.

Oklahoma Background, Policy History, and RPS Strength

Oklahoma is the twenty-eighth most populous US state, with a population of approximately four million—similar in population to Oregon and South Carolina. Oklahoma is also a very rural state. About a third of its population resides in the counties associated with its two largest cities, Oklahoma City and Tulsa. The other two-thirds live almost entirely in rural areas, with several counties home to fewer than ten inhabitants per square mile (US Census Bureau 2021c). Like in South Carolina, these politically conservative, rural areas are overrepresented in the state legislature and the state's current politics are extremely Republican-dominated. However, the Republican Party's takeover of Oklahoma's state government is a much more recent development than it was in South Carolina. The speed with which Oklahoma Republicans accelerated their dominance over the state legislature and executive branch during the 2010s is striking.

Table 4.1 shows the partisan composition of Oklahoma's state government from 1995–2021. While Republicans have controlled the House since

Table 4.1
Partisan control of Oklahoma's state government, 1995–2021

Year	Senate (Republicans/ Total Percentage)	House (Republicans/ Total Percentage)	Governor (Name (Party))
1995	11/48 (22.9)	33/101 (32.7)	Francis Keating (R)
1996	12/48 (25)	36/101 (35.6)	Francis Keating (R)
1997	12/48 (25)	36/101 (35.6)	Francis Keating (R)
1998	15/48 (31.3)	36/101 (35.6)	Francis Keating (R)
1999	15/48 (31.3)	36/101 (35.6)	Francis Keating (R)
2000	15/48 (31.3)	40/101 (39.6)	Francis Keating (R)
2001	15/48 (31.3)	40/101 (39.6)	Francis Keating (R)
2002	18/48 (37.5)	48/101 (47.5)	Francis Keating (R)
2003	20/48 (41.7)	48/101 (47.5)	Brad Henry (D)
2004	20/48 (41.7)	48/101 (47.5)	Brad Henry (D)
2005	22/48 (45.8)	57/101 (56.4)	Brad Henry (D)
2006	22/48 (45.8)	57/101 (56.4)	Brad Henry (D)
2007	24/48 (50)	57/101 (56.4)	Brad Henry (D)
2008	24/48 (50)	57/101 (56.4)	Brad Henry (D)
2009	26/48 (54.1)	61/101 (60.3)	Brad Henry (D)
2010	26/48 (54.1)	61/101 (60.3)	Brad Henry (D)
2011	32/48 (66.7)	70/101 (69.3)	Mary Fallin (R)
2012	31/48 (64.6)	68/101 (67.3)	Mary Fallin (R)
2013	36/48 (75)	72/101 (71.3)	Mary Fallin (R)
2014	36/48 (75)	72/101 (71.3)	Mary Fallin (R)
2015	40/48 (83.3)	72/101 (71.3)	Mary Fallin (R)
2016	39/48 (81.3)	71/101 (70.3)	Mary Fallin (R)
2017	42/48 (87.5)	75/101 (74.3)	Mary Fallin (R)
2018	40/48 (83.3)	72/101 (71.3)	Mary Fallin (R)
2019	39/48 (81.3)	77/101 (76.2)	Kevin Stitt (R)
2020	38/48 (79.2)	77/101 (76.2)	Kevin Stitt (R)
2021	38/48 (79.2)	82/101 (81.2)	Kevin Stitt (R)

Note: Middle columns of table show percentage of seats in each legislative chamber controlled by Republicans. Lighter shading indicates a simple majority for either party while darker shading indicates a two-thirds supermajority for either party. Source is Council of State Government's "Book of States."

2005, they have controlled the Senate only since 2009, with split party control of the Senate during the 2007 and 2008 legislative sessions. While Republican governors have been in power in Oklahoma since the arrival of Governor Mary Fallin in 2011 (who emerged victorious in the same 2010 midterm election that brought a Tea Party–driven Republican majority to the US Congress), it was a Democrat, Governor Brad Henry, who ran the state's executive branch from 2003 until then.

However, unlike South Carolina, Oklahoma is a "high" fossil fuel–producing state as shown in figure 2.2. Specifically, Oklahoma is an "oil and gas state," as described by many of those interviewed for this book. In 2010, the year Oklahoma adopted its voluntary RPS policy, and following the explosion in gas production enabled by the "fracking boom" (also known as the "shale revolution") of the mid-2000s, Oklahoma's combined oil and gas production was more than 2,474 trillion Btu. This was the third highest after only Texas and Wyoming, and Oklahoma's oil and gas production was higher than Texas' on a per capita basis (US Energy Information Administration 2023).

Until the mid-2000s, most of Oklahoma's electricity generation came from coal-fired power plants; beginning in 2009, gas-fired electricity generation overtook coal-fired generation. Around this same time, wind generation began to take off in the state and has been growing ever since. In 2002, 96.3 percent of Oklahoma's electricity was generated from fossil fuels and none was generated from wind. By 2021, 55.9 percent was generated from fossil fuels (primarily gas), and some 40 percent came from wind alone (US Energy Information Administration 2022). While it is unlikely that even a voluntary RPS would have been politically feasible in Oklahoma were it not for the state's enormous wind energy potential (Stadler et al. 2017), given the toothless nature of that 2010 law, state and federal tax credits were likely far more helpful in spurring the wind industry's impressive growth in Oklahoma (author's interviews, multiple).

Nevertheless, the dominance of the oil and gas industry in Oklahoman politics and culture is evident not only in the quantitative data from the US Energy Information Administration but also in the qualitative data gathered from a month of fieldwork in the state. As figure 4.1 shows, a model oil drilling rig is prominently displayed directly in front of the state's capitol building (there is no renewable energy equipment, such as a wind turbine,

Figure 4.1
A model oil drilling rig displayed in front of Oklahoma State Capitol. *Note*: Photographed by Joshua Basseches, 3/9/23.

to rival it). As figure 4.2 shows, a major office building in Oklahoma City is known as the "The Oil Center."

Two major oil and gas firms in particular, Continental Resources and Chesapeake Energy, are both headquartered in Oklahoma City. The executive chairman of the former, Harold G. Hamm, has an estimated net worth of nearly $50 billion and is the namesake of local hospitals, museums, and other public spaces in Oklahoma. Aubrey McClendon, the former CEO of the latter, was another towering figure in Oklahoma politics. Tulsa, Oklahoma's second most populous city after Oklahoma City, is known as the

Figure 4.2
The Oil Center in Oklahoma City. *Note*: Photographed by Joshua Basseches, 3/9/23.

"Oil Capital of the World," according to the American Oil and Gas Historical Society (American Oil and Gas Historical Society 2024). One of the city's landmarks (which is also the official state monument) is a seventy-six-foot-tall statue known as the "Golden Driller," which was built by the Mid-Continental Supply Company for the International Petroleum Exposition, which the city hosted in 1953 (Travel Oklahoma 2024).

Regarding the general influence of this industry in the state's public policymaking, a former top Oklahoma energy official stated point blank, "There is nothing stronger in the State of Oklahoma than the oil and gas lobby" (author's interview, 2/27/23). A policy advocate who regularly intervenes in dockets at the Oklahoma Corporation Commission (OCC), Oklahoma's

Figure 4.2
(continued)

PUC, said of the oil and gas industry: "They're the most influential industry in the state by far, and it's the backbone of the economy here" (author's interview, 3/16/23). When asked about the political strength of the oil and gas industry in several specific policy areas related to energy and electricity, a wind energy advocate described the industry as "strong in everything, insanely [strong]" (author's interview, 2/16/23). Perhaps the most illustrative quote on this question came from a former Oklahoma state legislator, who served in both the House and Senate for a total of twelve years:

> First of all, the gas industry runs the state capital of Oklahoma. I mean, I literally heard ["name of oil and gas company"] lobbyists say at a conference that ["name of oil and gas company"] had the best legislature money could buy in Oklahoma.

> I mean, pretty much the oil and gas industry got what they wanted. . . . That's just how it was. They threw around enough money to campaigns and things that they could typically get what they wanted. And that's probably still true today. . . . You know, [oil and gas lobbyists] come in here a lot, and have dinners most nights for members, and, of course, tickets to the Thunder[1] and other things like that. So they have a lot of influence at the Capitol. (author's interview, 3/1/23)

By comparison, "green business" actors (Basseches 2024) have been politically weak in Oklahoma. This has started to change only very recently, given the explosion in in-state wind generation mentioned previously. But interviews with political insiders from the early days (leading up to the enactment of the voluntary RPS policy in 2010) suggest only a tiny lobbying presence, certainly tiny relative to oil and gas. When asked of the 2010 RPS legislation whether it was something that wind industry lobbyists were asking for, the chief legislative sponsor of the legislation replied, "Nope. Did it on my own, literally on my own because I liked the idea" (author's interview, 2/21/23). As a former executive branch official recalled: "I don't believe there was anything quite as organized as the Oklahoma Wind Association in those early days . . . In fact, the entity that I recall being most involved . . . on behalf of wind power was a local—well, they were a family actually, that produced small wind turbines . . . I don't remember any larger group representing wind or renewables" (author's interview, 2/24/23).

A former state legislator noted: "The wind industry at the time was not powerful. It probably is more so now that you see more of the turbines being put up, particularly in the West [Western part of the state]" (author's interview, 3/1/23). Evidently, the political representation of renewable energy industries paled in comparison to that of the fossil fuel industry, especially oil and gas, in the years leading up to adoption of Oklahoma's voluntary RPS. The owner of the family wind business mentioned above recalled: "When I was involved, there wasn't much money in [wind]. Nobody [working in public policy advocacy] got paid. We didn't have any paid lobbyists. It was all volunteer stuff" (author's interview, 2/17/23).

That said, the minimal political influence of "green business" actors (Basseches 2024) in Oklahoma even in the early days would have seemed large relative to that of *nonprofit* climate-focused environmental advocacy groups. Of the seven states examined in this book, the influence of environmental groups in Oklahoma was the most negligible, and it was striking the number of times, and the range of different ways, this point was made by those interviewed—regardless of the interviewee's particular vantage point on Oklahoma politics. A former utility lobbyist described such groups as having "never been particularly effective in Oklahoma" (author's interview, 3/2/23). The following exchange with a former state legislator (who happened to be a Democrat) is also illustrative:

> **Interviewer:** "What about environmental—to what degree do environmental groups here matter, like the Sierra Club or anything like that?"
>
> **Interviewee:** "They don't. . . . Sierra Club probably—if they did come out and take a position on a bill, it would probably do more harm than good."
>
> **Interviewer:** "And that was true regardless of Democrats or Republicans?"
>
> **Interviewee:** "Yes." (author's interview, 3/1/23)

A legislative staffer who worked on HB 3028, the voluntary RPS bill that became law in 2010, recalled that, even though "Sierra Club [was] the main environmental group" active in Oklahoma, "to be completely honest with you, I'm not really aware of them having a significant presence at the Capitol" (author's interview, 2/27/23). When asked if they had any involvement with HB 3028, he replied, "I don't recall visiting with them, no" (author's interview, 2/27/23).

Indeed, the Oklahoma chapter of the Sierra Club, which numerous interviews and the archival record suggest was the *only* climate-focused environmental public interest group lobbying the state legislature, was seen as "radical" and "fringe" rather than a mainstream player in state politics. This exchange with a wind energy lobbyist is illustrative:

> **Interviewer:** "To what extent are there environmental non-profit or environmental NGOs that are involved?"
>
> **Interviewee:** "Well, Sierra Club is here. But . . . they're not effective. They're wide-eyed radicals." (author's interview, 2/17/23)

A different clean energy lobbyist similarly said: "The Sierra Club has a presence, but the Sierra Club comes with its own strings [attached], unfortunately. . . . They do have a reputation. Some baggage comes with it" (author's interview, 2/16/23). When asked to explain what she meant by "baggage," this lobbyist explained that the group is viewed by Oklahoma policymakers as having a certain ideological bend and trying to import policy ideas "from California" that just aren't palatable in Oklahoma (author's interview, 2/16/23).

When it comes to attempting to influence regulatory policy made at the OCC, the Sierra Club is not even a reliable *presence*. There is a different public interest group, the "Oklahoma Ratepayers' Sustainability Council,"[2] which describes itself as both a residential ratepayer advocate *and* a sustainable energy advocate, which works primarily at the OCC because the legislature is "too toxic" (author's interview, 3/16/23). The sole policy advocate of this organization said of the Sierra Club: "They don't engage at the Corporation Commission anymore. They did off and on through the years, but a couple of years ago they more or less stopped. And there's no other environmental organization, so it's just [us]" (author's interview, 3/16/23).

As a final illustration of just how politically weak the environmental lobby has been in Oklahoma, consider a former top state energy official's explanation of the decision to merge two state departments—Energy and Environment—under Governor Fallin's administration: "My sense is that you had a governor that had run under, 'I'm going to do a lot of Republican consolidation, make government more efficient.' And so, what she needed to do was to be able to say she had done some of those things. *Who the hell in Oklahoma cares about the environment?* I'm being silly, but, you know, let's merge those two together [emphasis added]" (author's interview, 2/27/23).

Two major vertically integrated IOUs dominate Oklahoma, serving well over 98 percent of the state's IOU customers over the period of 1995–2021: Oklahoma Gas and Electric (OG&E) and Public Service Company of Oklahoma (PSO). While both are publicly traded corporations, OG&E is a standalone company whereas PSO is a subsidiary of American Electric Power, a major energy parent company owning nine other regulated IOU subsidiaries operating in multiple states, plus nonregulated subsidiaries as well. American Electric Power bought PSO from its previous owner, Central and South West Corporation, in 2000 (The Oklahoman 2000). Not surprisingly given the rural geography of the state, a significant share of the state's customers are served not by IOUs but by rural electric cooperatives. During the period under study, about 66–69 percent were served by IOUs with most of the difference served by rural cooperatives; about 4–10 percent over the years were served by POUs. These POUs include mostly municipal utilities like the City of Edmond or the state-owned utility known as the Grand River Dam Authority (US Energy Information Administration 1995, 2001, 2008, 2021).

Although Oklahoma is a nonrestructured state that does not allow retail competition, it is unlike the other nonrestructured states in this book (Oregon and South Carolina) in that it is part of an RTO, the Southwest Power Pool (SPP). As a result, even though OG&E and PSO own significant generation assets of their own, a relatively greater share of electricity delivered to Oklahoma customers is generated by IPPs, approximately half of it. By comparison, in Oregon that number is less than 25 percent and in South Carolina it is only about 3 percent. That said, in 2021, approximately 95 percent of Oklahoma's *renewable* electricity was generated by these IPPs whereas nearly all the fossil-fueled generation came from facilities owned by OG&E or PSO (US Energy Information Administration 2021a).

OG&E owned major fossil-fueled power plant facilities, including the 617-MW gas plant, Rosebud (Oklahoma Corporation Commission 2008a), the 516-MW and 520-MW coal plants, Sooner 1 and Sooner 2 (Power Technology 2024), and the Seminole 1, 2, and 3 gas-fired units with a total capacity of 1,522-MW (Oklahoma Gas and Electric 2024). PSO, for its part, owned fossil-fueled power plants such as the 465-MW coal plant, Northeastern 3, the 422-MW and 434-MW gas plants, Northeastern 1 and 2, and the gas plant units, Riverside 1 and 2, with a total capacity of 896-MW (Public Service Company of Oklahoma 2018).

From a political economy perspective, this would suggest closer alignment between the IOUs and the fossil fuel industry on renewable energy mandates (opposing them) than we would expect to see in some other states, where IOUs themselves own significant renewable generation assets. That said, as one interviewee pointed out, the IOUs, which own the in-state transmission network, still stand to realize some profits from the proliferation of wind generation facilities, since they drive demand for transmission expansion, which the regulated IOUs can collect a return on for their shareholders (author's interview, 3/2/23).

Restructuring Era in Oklahoma

An often-forgotten part of Oklahoma's energy policy history is that the state *almost* restructured its electric utility sector in the late 1990s and early 2000s, but it ultimately failed to enact SB 220 in 2000, which would have greenlighted the restructuring provisions spelled out in earlier legislation (SB 500). SB 500 *did* become law in 1997, but required further legislative authorization for the restructuring provisions to take effect—authorization that never came. Even in 2001, with the enactment of SB 440, certain political forces in the state had not given up on omnibus utility restructuring legislation. SB 440 established the Electric Restructuring Advisory Committee and gave it the mandate to "continue the examination of electric industry restructuring issues" (Oklahoma Senate Bill 440, §4).

During this period, Oklahoma's electricity rates were certainly lower than the other states in this book that pursued restructuring legislation to completion at roughly the same time. As with other states, residential customers in Oklahoma (who did not enjoy the same quality of representation at the OCC as industrial customers did) paid higher rates than industrial customers. But both classes of customer paid significantly more in states like California, Pennsylvania and Massachusetts than in Oklahoma. Texas, which also pursued restructuring (see chapter 3), had *slightly* higher rates than Oklahoma. In 1997, for instance, Texas residential customers paid an average of 7.82 cents per kilowatt hour (compared with 6.63 cents in Oklahoma), and Texas industrial customers paid an average of 4.05 cents per kilowatt hour (compared with 3.63 cents in Oklahoma; US Energy Information Administration 1998).

Interestingly, and perhaps disturbingly, the archival record was missing some key pieces of documentation from that period about the Electric Restructuring Advisory Committee's deliberations and work. However, the records that *were* available, coupled with several interviews with key players that were involved at the time, tell an interesting and perhaps surprising story. Unlike in other states that went all the way forward with restructuring, where the effort was initiated by industrial customers and/or retail power marketers (e.g., Enron), in Oklahoma, the powerful, vertically integrated IOUs, OG&E and PSO, were the ones pushing most strongly *for* restructuring. This is because they saw it as a business opportunity to create a facade of competition while exercising market power privileges without sufficient regulatory guardrails. As one stakeholder put it, they were striving to go "from a regulated monopoly to a deregulated monopoly" (author's interview, 2/22/23). Unlike in Texas, where there were numerous out-of-state actors ready to compete with incumbent IOUs and sufficient market power protections enshrined in the statute, in Oklahoma, proposed, but ultimately not enacted, legislation would have created a situation in which incumbent IOUs would have become more powerful, not less (author's interviews, multiple).

A former utility regulator, in office at the time this restructuring legislation in Oklahoma was being considered, explained: "There was not the ability for competition. It would have been a 'deregulated monopoly.' I mean, the wires weren't opened up so that, you know, we could have a competitive marketplace for people to sell electricity. The wires were all at that point completely controlled by the investor-owned utilities, so it would have just made the investor-owned utilities a *deregulated monopoly* because they would control everything. And that was why I was opposed to it [emphasis added]" (author's interview, 3/6/23).

When I mentioned to this interviewee that other states' restructuring legislation contained market power provisions to protect against this sort of situation, this former utility regulator replied, "Sure, but that wasn't part of this legislation [in Oklahoma]. . . . It was being written by the utilities" (author's interview, 3/6/23). When asked where other interest groups were on the legislation, including the industrial customers (which, in other states, were the main drivers in favor of such legislation), this former utility regulator reiterated, "It was the utilities [that were the only ones pushing for it]" (author's interview, 3/6/23). A state legislator in office at the time confirmed, "It came from the electrical industry" (author's interview, 3/1/23). The former utility regulator continued: "It was gonna be a deregulated monopoly because there was market power that was not being offset by anything else. And I did not say that we couldn't do a restructuring bill, but the restructuring bill they were trying to cram down was not one that had the safeguards in it that could allow for a marketplace like that. . . . It was our job to protect the consumers and our industrial consumers and the homeowners" (author's interview, 3/6/23).

In the archival record, although many documents were missing, there was a single key document from that period, a 2001 "Interim Report Relating to Transmission Issues," authored by the Oklahoma Electric Restructuring Advisory Committee pursuant to SB 440. That SB 440 legislation had noted, in §4(H), "The Advisory Committee shall be authorized to employ any legal counsel, independent consultants or other persons as necessary to assist the Advisory Committee in the performance of its duties" (Oklahoma Senate Bill 440). It appears from the report that the Committee elected to do so, as the cover letter was authored by an attorney by the name of "Tom Smith"[3] (Oklahoma Electric Restructuring Advisory Committee 2001).

When I asked the former regulator who this person was, after having been unsuccessful in trying to track him down myself to interview him, the former regulator told me, "'Tom Smith'[4] is a former utility executive, just so you know" (author's interview, 3/6/23).

This digression is an important one for thinking about the political influence of IOUs, as I do at greater length elsewhere (see Basseches 2024, and chapter 5 of this volume). Others' research, too, has found IOUs' influence to be unparalleled (e.g., Culhane et al. 2021; Hirsh 1999; Peskoe 2023; Stokes 2020). Readers will recall that chapter 1 opened with an anecdote from an IOU lobbyist in Oregon about how his objective was not necessarily to *kill* renewable energy legislation but rather to *shape* it to his employer's advantage. In cases like Texas (and, as we will soon see, also Pennsylvania), IOUs were not always *enthusiastic* about restructuring, but they made it work well for themselves nevertheless, through generous provisions allowing them to recoup so-called stranded costs from ratepayers. Others have argued that these stranded cost recovery processes constitute unjustified windfall profits (Lupica 2000).

At the very least, these provisions "bought off" IOUs, neutralizing their opposition to the restructuring legislation. And then, in the Oklahoma case, we learn that IOUs actually attempted (although they were unsuccessful) to *promote* restructuring legislation as a vehicle to better themselves financially (despite already being vertically integrated monopolies) by achieving market power in a nascent RTO and reducing the quality of regulation they would be subject to.

The IOUs' inability to ultimately *enact* restructuring in Oklahoma should not be interpreted as a sign of their political weakness. On the contrary, my interview with a former Oklahoma IOU lobbyist suggests that the IOUs at the time viewed whatever the ultimate fate of restructuring efforts might be to be something of a "win-win," stating, "PSO, OG&E had really significant leverage in at the legislature. . . . We were able to basically feel pretty comfortable [either way]" (author's interview, 3/2/23). The conversation continued:

> **Interviewer:** "So, I mean, either scenario wouldn't have been bad? It's just that with one scenario, they [his company at the time] had been planning for it, whereas the other one was sort of the status quo, which they were happy with anyway?"
>
> **Interviewee:** "That's a great description." (author's interview, 3/2/23)

As for why restructuring never ultimately came to fruition in Oklahoma, it was likely due to a combination of factors. First, the industrial customers' trade association, which had some overlap with the oil and gas industry and was quite politically influential, ultimately opposed the legislation once it was unveiled (author's interviews, 2/22/23, 3/2/23, 3/6/23). Second, the Attorney General's office, which was responsible for advocating for all the state's ratepayers including residential and had a key seat on the Electric Restructuring Advisory Committee (Oklahoma Senate Bill 440), opposed the legislation, deeming it risky, especially given that electricity rates were already low relative to other parts of the country (author's interview, 2/24/23). And finally, at around the same time as this was all pending in Oklahoma, the California electricity crisis was unfolding on the West coast, and many were blaming it on California's restructuring legislation, which had been enacted a few years earlier (author's interviews, multiple). A former staffer in the Oklahoma Attorney General's office explained: "The AG [Attorney General], from my memory, was for the most part opposed to deregulation [restructuring]. And I think a lot of that had to do with the California experience and some of the horrors that, you know, came from that. . . . It just got to a point where it faded away with no action and I think it was because there was just not an appetite for deregulation at that time, based on what was going on around the country" (author's interview, 2/24/23). A former Oklahoma IOU lobbyist recalled: "When issues started happening in . . . California, other states, where they were having all kinds of major . . . problems. And basically . . . the Oklahoma legislature got cold feet" (author's interview, 3/2/23).

SB 440, which *did* become law, included state production tax credit (PTC) provisions, which were critical in helping the wind energy industry get off the ground, given the state's enormous and at the time completely underutilized wind energy potential (Stadler et al. 2017). Beginning in tax year 2002, "electricity generated by zero-emission facilities" (defined to include wind, moving water, sun, and geothermal energy) would be eligible for a tax credit of $0.0075 cents per kilowatt-hour, an amount that would ramp down to $0.0025 cents per kilowatt-hour until it would eventually expire ten years later, in 2012 (Oklahoma Senate Bill 440, §5).

Not surprisingly, this state PTC was extremely helpful at a time when the wind industry in Oklahoma was struggling to get off the ground. Both of the two Oklahoma wind industry executives interviewed for this project attested to its significance in helping their (at the time) small businesses personally (author's interviews, 2/17/23, 2/27/23). Interestingly, in those early days when wind development was in its infancy, the state's oil and gas lobby did not even view it as much of a threat and did not mobilize to prevent it. As a wind lobbyist recalled: "They didn't seem to care that much. I don't think they really saw it as a threat. I think by the time they recognized it as a threat, the horse was already out of the barn" (author's interview, 2/17/23).

He also suggested that perhaps the oil and gas industry was distracted by a significant battle brewing around the same time between gas and coal, with the gas industry trying to stop the construction of a proposed coal-fired plant by OG&E, which they ultimately succeeded in doing (author's interview, 2/17/23). But in those early days, as one lobbyist put it, "Oil and gas, I mean, they felt that it was a tax benefit that was more of a positive situation for renewables than a negative for oil and gas" (author's interview, 2/27/23), again suggesting that the oil and gas industry underestimated the threat posed by the wind industry's growth. Perhaps this was not naivete on the oil and gas industry's part as much as it was a genuine feeling that wind had *a role to play* in the overall electricity mix: "If you look at the technology, and if you look at the load patterns and all that, natural gas, especially combustion turbines, are well-suited to balancing variable production of renewables" (author's interview, 3/16/23).

However, it was not long before the oil and gas industry came to see the wind industry as a competitive threat rather than a complement. This was evidenced in the fight over HB 2469, legislation introduced in 2006 by a Democratic state representative from the wind-rich western part of the state. This legislation, which ultimately died in conference committee (presumably due to the mobilization of the oil and gas industry), would have extended the state PTC for wind through 2016 (Oklahoma House Bill 2469). The following comments illustrate this turn against wind's growing economic and political fortunes by the oil and gas lobby:

> But I think, over time, they [oil and gas] began to see how large wind was becoming, that it was eating into sales of natural gas. So they got pretty aggressive about opposing and trying to block wind power.
>
> —Intervenor at the OCC (author's interview, 3/16/23)

> [By the mid-2000s] there was quite a push to go ahead and do away with it [the state PTC for wind]. . . . I think it [the tax credit] ended because of the oil and gas industry lobby position down at the Capitol.
>
> —Renewable energy advocate (author's interview, 2/27/23)

> There was a huge competitive thing between wind, you know, being a source of electricity versus natural gas being that source of electricity.
>
> —Former OCC commissioner (author's interview, 3/6/23)

Of course, those without a direct stake in the oil and gas industry's profitability saw things differently. For them, public policy had traditionally privileged the oil and gas industry in Oklahoma, and the state PTC was an important tool to level the playing field, and develop a new, promising sector of the state's economy. A former lobbyist first characterized the oil and gas industry's position before offering a rebuttal:

> I can tell you from my experience representing [a wind industry client[5]] that there was this kind of attitude by several of the legislators, which is, well, you're not an infant industry anymore. You're grown now and you don't need the [tax] credits. You shouldn't need the [tax] credits. Now, my response to that was, well, when is the *oil and gas industry* going to be considered grown? Because they still receive lots of credits in different forms and they've been around 100 years in Oklahoma. So, you know, and I think it's just that tension between oil and gas industry and wind and other forms of renewable [energy] that I think is always going to exist in Oklahoma [emphasis in original]. (author's interview, 2/24/23).

A former utility regulator stated plainly: "At that point, the oil and gas industry in Oklahoma was trying to eliminate the production tax credit in Oklahoma and trying to put new taxes on the wind industry" (author's interview, 3/6/23). And a renewable energy lobbyist pointed out resentfully, "They peeled back or idled most of the tax credits, except for the oil and gas [ones], which they of course kept on" (author's interview, 2/17/23).

Oklahoma's Voluntary RPS

It is in this context that Oklahoma policymakers began exploring a mandate for renewable energy. This exploration eventually culminated in the state's purely voluntary and aspirational RPS policy, enacted by the state legislature in 2010. Table 4.2 shows how Oklahoma compares with Pennsylvania, the other state analyzed in this chapter, when it comes to the

Table 4.2
Oklahoma and Pennsylvania: Explanatory factors

	Partisanship	Energy Economy	Utility Sector Structure
Oklahoma	"Red," but only in recent years	High fossil fuel production	Vertically integrated
Pennsylvania	"Red," but barely	High fossil fuel production	Restructured

three major explanatory factors this book explores: partisanship, energy economy, and utility sector structure.

The effort to support renewable technologies through state policy, beyond the state PTC, began with a docket that was opened at the OCC "to investigate the potential of renewables for the production of electricity for Oklahoma," which specified a variety of technologies including wind, solar, hydropower, biomass and heat recovery. The aim was to gather comments from "interested parties" on the potential of these technologies to constitute a greater share of the state's electricity mix (Oklahoma Corporation Commission 2008). Respondents included OG&E, the Attorney General's office, the industrial customers' trade association, Walmart, and others. Since a mandate was not explicitly on the table in this docket, most of the comments filed were short and nonspecific, expressing interest in *exploring* the role renewables might play but, at the same time, expressing concern that no portfolio shifts increase the burden on ratepayers (e.g., Oklahoma Gas and Electric 2008; Oklahoma Attorney General 2008; Oklahoma Industrial Energy Consumers 2008; Walmart 2008).

During the 2010 legislative session, the Republican Speaker of the House sponsored HB 3028, the state's voluntary RPS, which was intentionally titled the "Oklahoma Energy Security Act" as opposed to any title explicitly highlighting renewable energy. It was signed into law by Democratic Governor Brad Henry on May 27, 2010. Section 4 states, in part: "It is hereby declared the *intent* of the State of Oklahoma to increase the use of renewable energy in the state by setting a renewable energy standard that will serve as a *goal* that 15 percent of all installed capacity of electricity generation within the state by the year 2015 be generated from renewable energy sources [emphasis added]" (Oklahoma House Bill 3028, §4).

It is important to note the explicitly aspirational language of the legislative text. "Renewable energy sources" are defined as "wind, solar photovoltaic, hydropower, hydrogen, geothermal, [and] biomass, which . . . may include agricultural crops, wastes, and residues, wood, animal and other degradable organic wastes, municipal solid waste, and landfill gas" (Oklahoma House Bill 3028, §4).

Interestingly, and unlike other RPS statutes in other states, HB 3028 also contained a provision emphasizing the importance of natural gas, although it does not qualify for the 15 percent goal. This provision reads, in part:

> The Legislature declares that it is in the public interest to promote natural gas energy development in order to take advantage of our state's vast natural gas resources and its compatibility with wind energy. It is further declared the intent of the Legislature to promote natural gas energy development in this state by establishing a natural gas energy standard to increase the use of natural gas as an electricity generation fuel . . . The natural gas energy standard shall be achieved by declaring natural gas as the preferred choice of electric generation for new fossil fuel generating facilities in this state. (Oklahoma House Bill 3028, §5)

Clearly, Oklahoma's character as an "oil and gas state" permeates even its sole legislative effort to promote renewable energy through a voluntary RPS. HB 3028's sponsor explained:

> We defined natural gas at the time as 'alternative' [energy], even though, again, somebody could argue with that. But for the purpose of political and legislative semantics, we wanted to push all of the above [in terms of energy technologies]. So we had a natural gas push. I leased a natural gas car from the House. When I would go to meetings or speeches, my staff would drive me in a natural gas car. (author's interview, 2/21/23)

The toothless nature of HB 3028 was never in doubt, although there was a slight textual difference between the Senate-engrossed and House-engrossed versions (which were ultimately reconciled in a legislative conference committee, as is typical). The Senate version contained some slightly more stringent language directing the OCC to implement additional rules to effectuate the law. A former legislator speculated that this language may have been added by a relatively progressive Democratic state senator who represented Oklahoma City at the time but who I was unable to interview to confirm whether this speculation was accurate (author's interview, 2/23/23).

In any event, the final version of the bill, which became law, was more bare-bones, removing any doubt that this was a voluntary, not mandatory, RPS. As the bill's sponsor recalled:

> We purposely called it "goals" [as opposed to a mandate]. . . . I don't recall there being any [interest] groups opposing it. I think I can say with accuracy that there would be some members who would say, "Well, I don't know if my oil guys are really comfortable with it." . . . But nobody, to my knowledge and recollection, ever organized and opposed it. . . . I don't remember it being terribly controversial. I think once it got to the floor, I think it passed pretty easily. (author's interview, 2/21/23)

Part of the reason the bill was viewed as so innocuous, in addition to the toothless legislative language, is that the state was already well on track to meeting the relatively modest goal of 15 percent renewable electricity by 2015, thanks in part to the state's abundant wind energy potential, and the state PTC discussed earlier, which was still in effect when HB 3028 became law. Indeed, data from the US Energy Information Administration shows that by the time HB 3028 became law in 2010, already about 10 percent of the state's electricity came from qualifying sources. The 15 percent goal was met by 2013, with two years to spare, meaning compliance would have been achieved even if the RPS *had* been mandatory (US Energy Information Administration 2022). As a legislative staffer who worked on HB 3028 recalled, "That 15%, I remember the feeling that it was very achievable" (author's interview, 2/27/23).

Table 4.3 shows how Oklahoma's voluntary RPS compares with Pennsylvania's mandatory one. Neither state has an RPS policy that rivals those of leading climate policy states like California, Massachusetts, or Oregon.[6] But that is unsurprising, given the dependence of both Oklahoma's and Pennsylvania's economies on fossil fuel production. Still, Pennsylvania's RPS is clearly a more environmentally serious one than Oklahoma's. Later in the chapter, we will explore why.

Oklahoma Within-Case Analysis

Before turning to the Pennsylvania case, a summary of the preceding within-case analysis of Oklahoma is in order. The evidence from Oklahoma suggests that—unlike we saw in Chapter 3 with Oregon and South Carolina, where partisanship and ideology were the dominant factors distinguishing those states' RPS policy outcomes—in Oklahoma, energy economy

Table 4.3
Oklahoma and Pennsylvania: RPS outcomes (selected policy design features)

	Mandatory or Voluntary?	Nominal RPS Target (Barbose 2023)	Regulated Parties	Controversial Resource Eligibility
Oklahoma	Voluntary	N/A	None	Hydroelectricity and biomass
Pennsylvania	Mandatory	<25%	Electric distribution companies (EDCs) and Electric generation suppliers (EGSs)	Hydroelectricity and biomass included in "Tier 1"; "Tier 2" includes waste coal and other controversial technologies

mattered more. Not only was Oklahoma much later than South Carolina in becoming a "red state" in the first place, with Republican dominance of Oklahoma state government only becoming established at roughly the same period that the Energy Security Act (HB 3028) was passed, but also, interviews suggest that partisanship and ideology were not major factors in the politics of this voluntary RPS legislation in Oklahoma.

As a former Oklahoma Speaker of the House stated: "I think it [your views on the Energy Security Act] depended on where you [as a legislator] were from . . . I don't think it separated on party lines" (author's interview, 2/21/23). Two other former state legislators, one a Democrat and the other a Republican, had the following to say about the role of partisanship:

> [On energy policy] It didn't matter which party was in control.
>
> —Former Democratic state legislator (author's interview, 3/1/23)

> I [as a Republican] always had an interest in wind energy because it was something that I knew would be good for Western Oklahoma because of the investment, because of the jobs.
>
> —Former Republican state legislator (author's interview, 2/23/23)

And a former top state energy official in the executive branch described the Energy Security Act as "very bipartisan" (author's interview, 2/27/23).

Even though the Tea Party had reached its height in 2010 when HB 3028 (the Energy Security Act) became law, it was not the Republican Party nor any political ideology associated with it that played the largest role in explaining its deliberately toothless nature. Rather, it was the nature of the state's energy economy—being an "oil and gas" state—that was far more salient. As a wind energy advocate explained: "The RPS was never gonna gain any type of teeth, strictly because oil and gas was just vehemently against it" (author's interview, 2/27/23).

Also against it, though perhaps less vehemently, were the state's vertically integrated IOUs. The relative importance of that will be discussed in conjunction with a comparative analysis that includes the Pennsylvania case, to which we now turn.

Pennsylvania Background, Policy History, and RPS Strength

Pennsylvania is the fifth most populous US state, with a population of about thirteen million—making it the third largest state analyzed in this book, after California and Texas. Pennsylvania's geopolitics are characterized by its two major cities, Philadelphia and Pittsburgh, and vast amounts of rural area in between. Philadelphia County and Allegheny County, home to these two cities respectively, comprise 22 percent of the state's total population, but when you add in the Philadelphia "collar counties" that number nearly doubles, to about 42 percent (US Census Bureau 2021a). A lobbyist invoked a locally famous quote by James Carville[7] to describe the state's geopolitics: "Well, you know what James Carville said about Pennsylvania? . . . He said, 'Well . . . you got Philadelphia and you got Pittsburgh, and then you have Alabama in between'" (author's interview, 5/15/23).

Another interviewee, a former Pennsylvania PUC commissioner, described it this way:

> The state of Pennsylvania, in a kind of simplistic way, looks a lot like the country. On the east coast of Pennsylvania, you have Philadelphia. Very predominantly Democratic controlled. Democratic everything all along in that whole area up and down the Eastern side of the state. In the Western part of the state, you have Pittsburgh, which was at the time [late 1990s and early 2000s] largely Democratic. Now it's more purple . . . And the "T," which is between it and across the whole northern tier, is rural Republican. (author's interview, 4/11/23).

Table 4.4 shows the partisan composition of Pennsylvania's state government from 1995–2021. Based on the operationalization of partisanship over time used in this book (see figure 2.1), the state would be considered a "red state," but barely. In reality, Pennsylvania is more of a "purple

Table 4.4
Partisan control of Pennsylvania's state government, 1995–2021

Year	Senate (Republicans/ Total Percentage)	House (Republicans/ Total Percentage)	Governor (Name (Party))
1995	26/50 (52)	98/203 (48.3)	Tom Ridge (R)
1996	27/50 (54)	102/203 (50.2)	Tom Ridge (R)
1997	27/50 (54)	102/203 (50.2)	Tom Ridge (R)
1998	30/50 (60)	104/203 (51.2)	Tom Ridge (R)
1999	30/50 (60)	104/203 (51.2)	Tom Ridge (R)
2000	30/50 (60)	103/203 (50.7)	Tom Ridge (R)
2001	30/50 (60)	103/203 (50.7)	Mark Schweiker (R)
2002	29/50 (58)	104/203 (51.2)	Mark Schweiker (R)
2003	28/50 (56)	108/203 (53.2)	Ed Rendell (D)
2004	29/50 (58)	108/203 (53.2)	Ed Rendell (D)
2005	29/50 (58)	110/203 (54.2)	Ed Rendell (D)
2006	30/50 (60)	110/203 (54.2)	Ed Rendell (D)
2007	29/50 (58)	101/203 (49.8)	Ed Rendell (D)
2008	29/50 (58)	101/203 (49.8)	Ed Rendell (D)
2009	29/50 (58)	99/203 (48.8)	Ed Rendell (D)
2010	30/50 (6)	97/203 (47.8)	Ed Rendell (D)
2011	30/50 (60)	112/203 (55.2)	Tom Corbett (R)
2012	30/50 (60)	110/203 (54.2)	Tom Corbett (R)
2013	27/50 (54)	111/203 (54.7)	Tom Corbett (R)
2014	26/50 (52)	110/203 (54.2)	Tom Corbett (R)
2015	30/50 (60)	119/203 (58.6)	Tom Wolf (D)
2016	31/50 (62)	119/203 (58.6)	Tom Wolf (D)
2017	34/50 (68)	121/203 (59.6)	Tom Wolf (D)
2018	34/50 (68)	120/203 (59.1)	Tom Wolf (D)
2019	28/50 (56)	110/203 (54.2)	Tom Wolf (D)
2020	28/50 (56)	110/203 (54.2)	Tom Wolf (D)
2021	27/50 (54)	111/203 (54.7)	Tom Wolf (D)

Note: Middle columns of table show percentage of seats in each legislative chamber controlled by Republicans. Lighter shading indicates a simple majority for either party while darker shading indicates a two-thirds supermajority for either party. Source is Council of State Government's "Book of States."

state." While Republicans controlled the state Senate for this entire period (including a supermajority in that chamber in 2017–2018), the House was controlled by Democrats in 1995 as well as from 2007–2010. Furthermore, control of the executive branch has tended to oscillate between Republicans and Democrats throughout this period, and at the time of this writing, Pennsylvania's governor, Josh Shapiro, is a Democrat.

Thus, if looking over the course of the entire period, neither Oklahoma nor Pennsylvania is as "ruby red" of a state as South Carolina or Texas. While Oklahoma has been extremely Republican-dominated post-2010, it was much "bluer" before then, and while Pennsylvania's legislature has more often been controlled by Republicans than Democrats, the margins have generally been thin, and the legislature and governorship are generally in play for both parties during most election cycles.

Another similarity between Oklahoma and Pennsylvania is that they are both extremely "high" fossil fuel–producing states, as shown in figure 2.2. While Oklahoma is an "oil and gas state," Pennsylvania is a "coal and gas state." Pennsylvania has a long history of coal mining, and even has a county—one of many where such mining activity takes place—called Carbon County (Pennsylvania Department of Environmental Protection 2024). More recently, the advent of fracking in the Marcellus Shale regions of northern and western Pennsylvania has made the state a natural gas–producing powerhouse as well (Jerolmack and Walker 2018).

By the numbers, in terms of the 1995–2021 period under study, Pennsylvania's coal production peaked in 1998 at 2,133.5 trillion Btu. That year Pennsylvania was the fourth largest coal-producing state in absolute terms, after Wyoming, West Virginia, and Kentucky. Per capita, it was ninth. Pennsylvania's natural gas production began climbing exponentially following the discovery of fracking in the mid-2000s, reaching its peak in 2021 at 8,057.5 trillion Btu. In 2021, Pennsylvania's gas production was second only to Texas's. On a per capita basis, Pennsylvania surpassed even Texas (US Energy Information Administration 2023).

In terms of electricity generation, coal-fired and gas-fired power plants have historically dominated, together providing an average of 60 percent of in-state electricity generation during the period under study.[8] Of that roughly 60 percent, much more came from coal plants than gas plants. It was not until 2016 that gas-fired generation overtook coal-fired generation, producing more than sixty-eight million megawatt hours of electricity that year compared to coal's roughly fifty-five million (US Energy Information Administration 2022).

Renewable electricity generation (excluding nuclear), by contrast, has constituted a tiny share of Pennsylvania's total generation. Hydroelectricity produced no more than 2 percent of the state's total in any of the twenty-seven years under study. Wind was not part of the state's electricity mix at all until 2000. Since then, its share has grown, but only to constitute about 1.5 percent of the state's total electricity generation. And solar has perhaps fared the worst in Pennsylvania, not contributing to the state's grid at all until 2008, after which remarkably—even as of the year 2021—it contributed to less than 1 percent. This is comparable to the share of in-state solar generation in Oklahoma in that same year, but a significantly lower share than Oregon and South Carolina, where solar in both represented about 2 percent. By comparison, in California that same year solar represented 18 percent (US Energy Information Administration 2022).

Like Oklahoma, the influence of the fossil fuel industry in shaping Pennsylvania's energy policy over the years was evident in the qualitative data collected during intensive fieldwork in the state. As a former Pennsylvania PUC commissioner put it: "In Pennsylvania, all the political power was with the coal industry," noting that "the coal industry also worked through the Pennsylvania Chamber of Commerce . . . so they were basically the voice of the Coal Association and the gas producers" (author's interview, 5/5/23). This individual also identified the "Coal Association" as the single most significant opponent to the state's RPS policy when first proposed (author's interview, 5/5/23). Others independently delivered similar assessments. For example, a former official in the Pennsylvania Department of Environmental Protection (PDEP) said, "The Coal Association was like crusaders. They were absolutely against [the RPS proposal]" (author's interview, 5/16/23). And a professional environmental advocate recalled: "We were well aware that Pennsylvania was and is seen as a fossil fuel state. So, we didn't see sort of an easy path for renewables, you know, to get support from the state" (author's interview, 4/25/23).

Although coal industry forces were described as more politically influential and more staunchly opposed to the RPS proposal than gas industry forces,[9] the latter was also influential—especially post–Marcellus Shale boom—and contributed to a general "fossil fuel culture" amongst members of the state's political class. A utility lobbyist noted, "Pennsylvania is awash in natural gas, and we have the Marcellus Shale fields. . . . We used to get most of our gas from the Gulf of Mexico. Ninety percent of it now comes from Pennsylvania" (author's interview, 5/4/23).

A "traditional business" (Basseches 2024) association lobbyist described the impact of fracking in the Marcellus Shale regions:

> It was only really discovered in 2005. Production only really started to ramp up in 2008. And you know, when I'm communicating with people, I put it in these terms: "This is a world-historic event." That sounds hyperbolic. It is not. Like, this discovery is as big as the opening of the Panama Canal. And, like, for example, so the Marcellus play is so huge, untold trillions of recoverable cubic feet of natural gas. Susquehanna County, Pennsylvania, in the far northeast along the New York state border, is the number one natural gas-producing county in America. (author's interview, 5/12/23).

By contrast, "green business" (Basseches 2024) is far less developed in Pennsylvania, and, therefore, politically weaker than in other states in this book, such as California and Massachusetts, certainly, but even also states like Texas or Oklahoma (nowadays) given their sizable wind industries. While in 2021 wind provided more than 20 percent of Texas's electricity generation and more than 40 percent of Oklahoma's, it provided only about 1.5 percent of Pennsylvania's, and as discussed earlier, solar was virtually absent from Pennsylvania's electricity mix until several years *after* Pennsylvania's RPS became law (US Energy Information Administration 2022).

Correspondingly, "green business" actors' influence over the Pennsylvania RPS and related state electricity policies was muted. As one policy advocate, who represented the interests of low-income electricity customers, put it, "At that time, there were hundreds and hundreds of lobbyists for the fossil fuel industry and the utility industry. There was one lobbyist in the entire state, "Claudia,"[10] and she represented both the energy efficiency industry and the solar industry" (author's interview, 4/26/23).

In an interview, that lonely lobbyist, "Claudia," recalled how difficult the job was, in particular discussing the process of attempting to negotiate differences among her various "green business" clients: "I spent most of my time saying, 'Guys, if you aren't united on this, we're not going to get anything done'" (author's interview, 5/16/23). Although widely regarded as skilled at her job, "Claudia" noted that after several failed legislative attempts to enhance the RPS in recent years, "I left lobbying because I didn't see a path forward to get anything else done" (author's interview,

5/16/23). When asked about renewable energy businesses' impact on the initial RPS legislation, a professional environmental interest group advocate explained: "They had not gotten off the ground. They had not gotten off the ground at all. That came a number of years later" (author's interview, 4/25/23).

However, in contrast with Oklahoma, Pennsylvania *did* have a robust landscape of environmental interest groups, which were active in state policymaking venues. Environmental groups such as the Clean Air Council, the Pennsylvania Environmental Council, PennFuture, PennEnvironment, Clean Water Action, Sierra Club, and others appeared regularly before state legislative committees and/or intervened in state PUC proceedings (e.g., Pennsylvania Public Utility Commission 1996; Pennsylvania House of Representatives Archives 2004; Pennsylvania State House of Representatives Archives 2009). Regarding proceedings related to restructuring at the PUC, a former commissioner recalled, "Environmental groups were there" (author's interview, 4/26/23), and this is reflected in the archival record as well (Pennsylvania Public Utility Commission 1996).

In particular, two major foundations, the Heinz Endowments and the Pew Charitable Trusts, were mentioned by multiple interviewees as supporting environmental policy work in the state and indeed these two joined forces in the late 1990s "to create a new environmental organization in Pennsylvania" (author's interview, 5/5/23) called PennFuture, which was especially active on the RPS (Pennsylvania State House of Representatives Archives 2010). However, as we will see, the influence of such groups over Pennsylvania's RPS policy was diluted by major divisions *within* the environmental community about whether to withhold support for the RPS legislation when it turned out to be less stringent than some advocates had hoped, especially on the question of resource eligibility.

Also in contrast to Oklahoma, residential and commercial ratepayers enjoyed greater political representation in Pennsylvania.[11] This was in large part because of dedicated state-funded agencies that regularly intervened in PUC dockets, whether rate cases or rulemakings. The Office of the Consumer Advocate (OCA) was established by the state legislature in 1976 for the purpose of representing ratepayers, with a particular focus on residential ratepayers. Twelve years later, the Office of the Small Business Advocate (OSBA) was created to do the same for commercial ratepayers. The leaders of both offices are appointed by the state's Attorney General[12] (author's interviews 4/7/23, 4/12/23; Pennsylvania Office of the Consumer Advocate 2024; Pennsylvania Office of the Small Business Advocate 2024). To give a sense of the Office of the Consumer Advocate's significant administrative capacity, in 2007, the office had thirty-seven full-time employees including sixteen attorneys (Pennsylvania State House of Representatives Archives 2007).

From 1979 to 2012, the period in which both restructuring and the most significant RPS policy developments in Pennsylvania took place, the Office of the Consumer Advocate was led by Irwin A. "Sonny" Popowsky, who is widely regarded as extremely effective, having (among his many other accomplishments) successfully argued a case before the US Supreme Court that reportedly saved ratepayers $50 million (Maykuth 2012). A colleague described Popowsky's approach to the negotiations leading up to the state's 1996 restructuring law:

> Sonny knew exactly, you know, which way to turn at every point and made exactly the right decision. And Sonny would dig in his heels and he would not budge if he was protecting something fundamental. And he would make it really clear, "This is the line I've drawn. I will not move from this line. You can talk to me until you're blue in the face, but I'm not changing." And people respected him. There's a massive amount of respect for him. (author's interview, 4/26/23).

A former state legislator involved in authoring Pennsylvania's restructuring law described Popowsky simply as "a smart guy" (author's interview, 5/9/23).

Although the Office of the Consumer Advocate was the most consistently involved in Pennsylvania's electricity policymaking on behalf of residential ratepayers, several other advocacy organizations—which were interest groups as opposed to official government agencies—also frequently testified before the legislature and/or the PUC. Some of these focused specifically on the needs of a subset of residential ratepayers, such as the elderly or those on fixed incomes and low-income ratepayers more generally. Such organizations included the American Association of Retired Persons (AARP),[13] the Pennsylvania Utility Law Project, the Energy Coordinating Agency, and several other local community development corporations (author's interview, 4/7/23; Pennsylvania Public Utility Commission 1995, 1996a).

Despite several mergers and acquisitions involving energy parent companies during the period under study, there have generally been eight major

IOUs in Pennsylvania, which serve 97 percent of the state's IOU customers and are listed in table 1.1. Philadelphia Electric Company (PECO) is now a subsidiary of Exelon Corporation. Pennsylvania Power Company (Penn Power),[14] Metropolitan Edison (Met-Ed), Pennsylvania Electric (Penelec), and West Penn Power are all now subsidiaries of First Energy Corporation. Pennsylvania Power and Light (PP&L), UGI Utilities (UGI), and Duquesne Power and Light belong to relatively smaller holding companies (US Energy Information Administration 1995, 2001, 2008, 2021).

On average over the period under study, IOUs together served about 87 percent of the state's electricity customers (residential, commercial, and industrial). Cooperatives served a relatively small percentage of the state's customers—on average about 3 percent—despite the state having many rural areas. Only a little over 1 percent (on average) of the state's customers were served by POUs such as municipal utilities. Given Pennsylvania's decision to allow retail competition as part of its restructuring law (more on that to come), a significant percentage of the state's customers over the years have been served by RPMs—about 9 percent on average between 1995 and 2021, though that percentage was 0 percent until retail competition took effect following the 1996 restructuring law (US Energy Information Administration 1995, 2001, 2008, 2021).

Restructuring in Pennsylvania

The origins of Pennsylvania's 1996 restructuring law were in some ways similar but in other ways different from those of restructuring in other states. Similar to other now-restructured states, Pennsylvania's average retail rates for electricity at the time were high. In 1996, the average residential customer in Pennsylvania paid 9.73 cents per kilowatt hour, higher than in any of the three states in this book that opted *not* to fully restructure (Oklahoma, Oregon, and South Carolina). Consistent with the pattern across all the states in this book, Pennsylvania's industrial customers paid less than its residential customers, but still significantly more (on average, 5.93 cents per kilowatt hour) than in any of the states that did not fully restructure (US Energy Information Administration 1997).

On the other hand, what made the Pennsylvania case unusual was the enormous *discrepancy* in rates, depending which IOU service territory a customer was located in. This was due to several factors including the degree to which the IOU had invested in (expensive) nuclear plants versus (less expensive, at the time) coal-fired plants. For instance, PECO had invested heavily in nuclear plants, and consequently, its residential rates were nearly 13 cents per kilowatt hour, "much higher than the Pennsylvania average" (Pennsylvania Public Utility Commission 1996a). By contrast, customers in neighboring PP&L's service territory were paying rates that were 50 percent lower than PECO's, and the rates of customers in the territories of IOUs serving other parts of the state were also relatively low (Pennsylvania State House of Representatives Archives 1995). The unfairness of this reality in which customers could be paying vastly different amounts for electricity simply by virtue of where in the state they were physically located was a driving narrative behind Pennsylvania's restructuring effort (Pennsylvania Public Utility Commission 1996b).

On April 14, 1994, the PUC opened an investigatory docket, I-00940032, to investigate "the role of competition in the Pennsylvania electric industry" (Pennsylvania Public Utility Commission 1995a). This docket included reams of testimony and exhibits from IOUs; nonutility generators; RPMs like Enron, which were eager to establish a market in Pennsylvania; consumer advocates; environmental advocates; labor unions; local elected officials; and others. At the center of the docket were more than 1,400 pages of transcribed oral testimony from more than half a dozen PUC hearings held in Harrisburg, Philadelphia, and Pittsburgh between December 1995 and March 1996. The hearings consisted of the five Pennsylvania PUC commissioners at the time questioning witnesses representing the multitude of stakeholders involved on their preferences for what a restructuring law would look like, and interrogating the bases of these preferences (Pennsylvania Public Utility Commission 1995, 1995b, 1996, 1996a, 1996b, 1996c, 1996d).

The chairman of the PUC at the time recalls sitting down with Governor Ridge to receive his blessing for efforts to restructure the electric utility industry in Pennsylvania:

> I went to the governor and I said, "Governor, I'd like to try to get a competition bill through here, to try to negotiate a way that we could restructure the electric industry and become more competitive." Because it didn't seem to me, with technological changes and what was going on at FERC, that there was any reason why electricity should be vertically-integrated. . . . So I said to the governor, "Can I have your authority to negotiate a bill to put through the legislature?" And he said, "Yeah, that would be fine." But he wasn't interested in getting into a back-down, drag-about fight with consumers, environment, utilities, labor. I mean, you know the list of players . . . And he said, "If you can negotiate something

> that's close to unanimous, I'd be interested." So, I had that authority. He made a speech, made me a head of the negotiating team, and we sat down and we negotiated a bill from scratch. (author's interview, 5/15/23).

Meanwhile, separately, key members of the state legislature were also interested in restructuring. One in particular, Republican Representative Frank "Chick" Tulli, had introduced HB 2537 on April 10, 1996, which aimed to restructure the industry. He recalled, "The Republican Leader [of the House of Representatives] . . . started the new [1995–1996 legislative] session by saying, 'We've got problems in Pennsylvania. Companies are leaving and companies are not coming to Pennsylvania [due to the high cost of electricity]' . . . We had to do something about high electricity costs" (author's interview, 5/9/23).

Tulli's bill would eventually be folded into HB 1509[15], the "Electricity Generation Customer Choice and Competition Act," which was a carefully negotiated legislative product whose language was largely informed by a report from the PUC, which was the culmination of the lengthy deliberations discussed previously. That bill, HB 1509, was voted out of the legislature at approximately 4:00 in the morning on November 25, 1996, after extensive floor debate but no successful amendments[16] (Pennsylvania House Journal 1996; Pennsylvania Senate Journal 1996). Governor Ridge signed HB 1509 into law on December 3, 1996.

Although the types of stakeholders with an interest in how the details of the restructuring law would work out were similar to those in other states, one big difference was the vastly divergent preferences and attitudes of Pennsylvania's major IOUs. Certain IOUs *welcomed* restructuring very early on, sensing financial opportunities if the policy were crafted in the "right" way. For example, a representative of Met-Ed and its sister company Penelec testified, "We believe that the time is right. We believe that this country is interested, this society is interested in choice. We believe that competition does inherently produce benefits . . . and we believe that it is the right thing to do" (Pennsylvania Public Utility Commission 1996b, 421).

He went on to express numerous caveats (or "principles"), most importantly that the PUC "establish the principle of recovery of stranded costs" (Pennsylvania Public Utility Commission 1996b, 421), but the overall tenor of his testimony was that restructuring was a welcomed opportunity. Similarly, PP&L stated that the company "supports choice of electricity suppliers for all customers" (Pennsylvania Public Utility Commission 1996b, 497), noting that the company viewed restructuring in Pennsylvania as "inevitable" for a number of reasons, one of which was that "existing rate differentials among utilities are dramatic" and that "the result has been a perception that the current structure is unfair" (Pennsylvania Public Utility Commission 1996b, 498).

By contrast, PECO—not coincidentally the IOU with the highest rates and with the most to lose in terms of stranded costs under a restructured system—had a very different view. A PECO representative testified, "You have heard a great deal about current rate disparities and many theoretical claims that retail competition will bring lower rates to customers, but when it comes to cold, hard facts and data to support those claims, these hearings have provided a reality check on the idea of dismantling the existing system and replacing it with an untested and undefined form of retail competition" (Pennsylvania Public Utility Commission 1996b, 529).

This testimony went on to warn of stranded costs for the company that could approach $7 billion (with the assumption being, of course, that these costs would need to be absorbed by ratepayers) and enormous loss of tax revenues to the state (Pennsylvania Public Utility Commission 1996b).

Furthermore, whereas in other states IOUs appeared to present a more united and collaborative front to policymakers, in Pennsylvania, animosity between companies was more readily apparent. For example, a representative of the IOU, West Penn Power, noted that his company had made "good decisions" (Pennsylvania Public Utility Commission 1996b, 379), implying that others hadn't, which had kept rates low for its customers. He went on to express concern that whatever formula policymakers devised to achieve stranded cost recovery might create an unfair playing field for some IOUs over others, explaining, "Subsidized stranded costs write-offs will allow high-cost utilities to sell their high-cost power at a subsidized price, and the likely targets of this subsidized power will be the best customers of the low-cost utilities [e.g. West Penn Power]" (Pennsylvania Public Utility Commission 1996b, 381).

When it came to the other (non-IOU) stakeholders, views on restructuring were also mixed. Industrial customers, represented by the Industrial Energy Consumers of Pennsylvania (IECPA), were strongly supportive of retail competition (as were their counterparts in other states) but were concerned that IOUs might be permitted to "over-recover" stranded costs at their members' expense. Residential customers, represented by the OCA and other groups for specific sub-categories such as low-income, were "cautiously neutral" and, in the case of a few outside groups, opposed—worried mostly about cost-shifting associated with stranded cost recovery but also exploitation of less sophisticated customers by RPMs and competitive suppliers. Thus, they emphasized the need for consumer education programs, universal service guarantees, and other protections. The OSBA on behalf of commercial customers echoed many of the OCA's concerns (Pennsylvania Public Utility Commission 1995).

A representative for Three Mile Island Alert, a citizens' nuclear watchdog group, expressed concern that restructuring would fail to hold accountable the IOUs that had made poor investment decisions (in nuclear power). In fact, this group expressed concern that these companies would use restructuring as an opportunity to be "bailed out" by ratepayers: "My concern is we're almost into an arena now of corporate socialism, where when things go well, they [IOUs] profit; and when things go bad, they [IOUs] come with their hands out to the Public Utility Commission. I think these are mature people who made decisions and they have to 'fess up' for a decision if it goes wrong" (Pennsylvania Public Utility Commission 1995, 92). In other words, most of the stakeholders that intervened as the PUC was developing its restructuring proposal were concerned not with what the ultimate structure of the utility sector would actually be but rather the question of "who pays" (Basseches 2024) the transition costs of reforms that might ultimately be in the public's interest.

It is worth noting that environmental interest groups also testified in the Pennsylvania restructuring proceedings. Like their counterparts in Texas who succeeded in introducing that state's RPS as part of the restructuring legislation, the Pennsylvania environmental groups wanted to ensure that the "market" for electricity generation that the restructuring legislation would create would be fair to renewable generation sources, taking into account the negative externalities of fossil-fueled generation, even if (at the time) it may have been cheaper (Pennsylvania Public Utility Commission 1995). Others believed renewables could never be cost-competitive and therefore favored traditional regulation as the best chance of preserving opportunities to mandate the addition of renewables. One such environmental advocate recalled: "We felt that it would hurt the development of alternative energies, you know, whereas if it's regulated [i.e. not restructured], it [renewable energy] can be mandated by the PUC and that type of stuff" (author's interview, 4/25/23).

Although, unlike Texas's 1999 restructuring law, Pennsylvania's 1996 law did *not* include in it an RPS,[17] so environmentalists were able to achieve some of their aims through public benefit charges that were negotiated at the PUC as each individual IOU filed its "restructuring plan." As one environmental advocate recalled: "We intervened in all the [PUC] cases and we got pots of money. . . . In each of the utility service areas there were funds set up to support, like, renewable energy projects" (author's interview, 4/13/23). However, they were "not that much money" relative to the billions in stranded costs (author's interview, 4/13/23).

The RPMs, for their part, were bullish on establishing retail choice as soon as possible in the Commonwealth of Pennsylvania. This was their business model, and they were eager to enter Pennsylvania's very large market and to compete with incumbent distribution utilities for customers. The most striking aspect of their testimony at the PUC was the *degree* to which they wanted competition *as soon as possible*, while others (even those generally supportive of restructuring) were urging cautiousness and slowness to make sure the state "got it right." By contrast, a representative from Enron stated, "How long can Pennsylvania tolerate electric rates that are among the highest in the country? How long can Pennsylvania afford to pay $1.5 billion a year more than the national average in electricity costs every year? How long will the interests of monopoly firms be put before the interests of taxpayers, employers and homeowners? How long will utilities be allowed to say no when their customers ask simply to be able to choose? The answer I hope to each of these questions is not long" (Pennsylvania Public Utility Commission 1996c, 976). Enron's testimony continued: "Getting it done, though, requires a proactive role, not a reactive one. You have to set standards and you have to establish deadlines or the job will never get done" (Pennsylvania Public Utility Commission 1996c, 977).

On stranded costs, knowing that even customers who opted for a competitive supplier or RPM would likely be required to absorb them, Enron was far more skeptical than the IOUs themselves (understandably, since the IOUs had so much more at stake). The Enron representative stated:

> The other argument [from IOUs] is we'll all go bankrupt if you make this move to competition . . . If you look at *Fortune* magazine last year, their ranking of the top 20 returns on revenues, the utility industry ranked fifth. Now, that's only one

data point, but to me it ought to prompt further examination of the allegations that utilities were not compensated for the possibility of a move to competition and they weren't allowed to earn rates of return that were concomitant with what they would see if they were in a competitive line of business. I'm asking you to look at these arguments very hard. (Pennsylvania Public Utility Commission 1996c, 981).

Under questioning from the PUC Chairman about whether a "transition period" would be necessary during which IOUs could recover their stranded costs without worrying about immediately losing their customer base, the Enron representative said, "no":

Chairman: "So you believe that we should move to customer choice tomorrow?"

Enron: "Or later today, if possible [laughter]." (Pennsylvania Public Utility Commission 1996c, 981)

Finally, preexisting nonutility generators (also known as IPPs) who had already entered into contracts[18] with incumbent utilities were also concerned that their existing long-term contracts with IOUs be protected, and/or that the full value of these contracts be included in the stranded cost calculation, with those stranded costs being recouped not only by the IOUs but also by the relevant IPPs. Even though, perhaps ironically, restructuring opened the industry to many more IPPs than had existed previously, those few IPPs that *were* preexisting saw some risk in the prospect of the transition.

An example of one specific subset of these IPPs was the Appalachian Region Independent Power Producers Association (ARIPPA), which is "a trade association comprised of thirteen qualifying facilities,[19] all of which are in operation in Pennsylvania, and all of which use waste coal as their fuel source" (Pennsylvania Public Utility Commission 1996c, 938). Their primary concern was stated plainly in their testimony before the PUC:

Our primary concern with this debate in which we feel somewhat caught up—it's not a debate of our making—is that our contracts be respected regardless of which direction this Commission or the nation moves in. Our contracts are really the only thing we have, and on that basis, as I said before, all of the projects were financed, constructed and continue operation. Without the contracts, financing would have never been provided . . . We feel very strongly that our contracts are protected legally and cannot be reopened regardless of the outcome of the debate on retail wheeling.[20] (Pennsylvania Public Utility Commission 1996c, 938–939).

Additionally, interviews suggest there was also concern that waste coal as a technology would not have been cost-competitive in an open market environment, in the absence of contracts or subsidies (author's interviews, multiple).

In the end, Pennsylvania's restructuring law, HB 1509, was carefully crafted behind closed doors, in negotiations led by the PUC commissioners with the input of certain key legislators and representatives of many (though probably not all) of the stakeholders whose views were just discussed. Not surprisingly, given both the complexity of the issue as well as the PUC's direct involvement in crafting the legislation, the legislation itself delegated enormous authority and discretion to the PUC to preside over many important policy details.

In particular, the PUC was given the authority to review and render judgment upon each individual IOU's "restructuring plan," which was to be filed with the PUC sometime between April 1 and September 30, 1997. The PUC was required to render judgment upon it within nine months of its filing date (Pennsylvania House Bill 1509, §2806(d-f)). These plans were required to include: the unbundling of rates into their generation, transmission and distribution components; a proposed competitive transition charge (CTC) requiring the calculation of a proposed value of all stranded costs periodized over the course of the company's transition period (more on that soon); a proposed universal service and energy conservation cost recovery mechanism[21]; procedures for ensuring direct access [to monopoly-owned transmission and distribution system] to all [competitive] licensed electric generation suppliers; a discussion of the impacts of the proposed plan on the utility's employees[22]; and revised tariffs and rate schedules implementing the above (Pennsylvania House Bill 1509, §2806(e)).

That said, despite many details left to the PUC in its decision-making on individual IOUs' restructuring plans, there were certain aspects of the restructuring policy that *were* spelled out in the statute ahead of time. For example, the IOUs were each required to unbundle customer charges for generation, transmission, and distribution (indeed, this was to be part of the aforementioned restructuring plan; Pennsylvania House Bill 1509, §2804(3)). It was established that, while generation supply would ultimately become deregulated and competitive, transmission and distribution would "continue to be regulated as a natural monopoly subject to the jurisdiction and active supervision of the commission [PUC]" (Pennsylvania House Bill 1509, §2802(16)). IOUs would have to separate themselves into separate affiliate entities, one for generation and the other for transmission and distribution. This unbundling is similar to in Texas, but unlike in Texas, there were not three separate levels of utility companies that were required to be created, only two: "electric distribution utilities" (transmission and distribution), which would own poles and wires and be responsible for providing universal service to end-users, and "electric generation suppliers" (generation), which would compete at the wholesale level to supply electricity to the grid.

Also unlike in Texas, companies providing transmission and distribution service were not statutorily *required* to divest from the generation business. Section 2804 stipulates: "The commission [PUC] may permit, *but shall not require*, an electric utility to divest itself of facilities or to reorganize its corporate structure [emphasis added]" (Pennsylvania House Bill 1509, §2804(5)). A former PUC commissioner recalled, "The utilities really put their foot down hard on that. They didn't want to be forced out of the generation business. Some elected to voluntarily [leave the generation business] shortly thereafter, but they didn't want the mandate [to do so]" (author's interview, 4/26/23).

As it turned out, eventually *all* of the IOUs *did* voluntarily exit the generation business, though they did so each at their own pace: "They all eventually voluntarily divested. It just took awhile" (author's interview, 4/26/23). A second former PUC commissioner confirmed: "After the fact, they all did [divest]. A hundred percent of it was divested. In the room [negotiating HB 1509], most of the utility companies said they wanted to keep their generation. They moved from wanting to keep it to selling it voluntarily" (author's interview, 4/28/23).

The choice to voluntarily divest was a business decision rather than a regulatory requirement. A former Pennsylvania IOU executive explained: "[Our company's] motive to spin off the [generation] business, which I . . . executed the spinoff, was more of a shareholder pressure maneuver than anything. There was no requirement, no other impetus, except that the shareholders of these companies that were trying to play in a competitive marketplace, and a regulated utility . . . It began to confuse the shareholders. Who are you? . . . Shareholder interests get split" (author's interview, 5/18/23).

Another pro-IOU provision was the guaranteed recovery of *all* stranded costs through one of multiple mechanisms, the most important of which (similar to in Texas) was the "nonbypassable competitive transition charge

(CTC)," which would be added to ratepayers' bills throughout the duration of the transition period, which could last up to nine years. In exchange, the other components of customers' bills would be frozen at their prerestructuring levels until the transition period for each IOU was complete and all stranded costs were recovered (Pennsylvania House Bill 1509, §2804(4)). The calculation of the total value of these stranded costs was left to each IOU in its restructuring plan, subject to PUC approval, using an IOU-friendly "just and reasonable" standard (Pennsylvania House Bill 1509, §2808).

An additional key component of the statute was the phasing in of retail choice for all electricity customers: "As of January 1, 2001, consistent with the commission's discretion under this section, all customers of electric distribution companies in this Commonwealth shall have the opportunity to purchase electricity from their choice of electric generation suppliers. The ultimate choice of the generation supplier is to rest with the consumer" (Pennsylvania House Bill 1509, §2806(a)).

To make the transition more manageable and administer correctives as needed, retail choice was "phased in": "As of January 1, 1999, a maximum of 33% of the peak load of each customer class shall have the opportunity for direct access; As of January 1, 2000, a maximum of 66% of the peak load of each customer class shall have the opportunity for direct access; As of January 1, 2001, all customers of electric distribution companies in this Commonwealth shall have the opportunity for direct access" (Pennsylvania House Bill 1509, §2806(b)(1–3)).

Similar to in Texas, non-IOU utilities such as POUs and co-ops were given the opportunity to "opt-in" to the new restructured, direct access regime, but they were not required to do so (Pennsylvania House Bill 1509, chapter 74).

Perhaps the two strongest indications of IOUs' influence over the restructuring legislation were (1) the inclusion of exceptions that IOUs could apply for, and (2) the inclusion of §2812, which provided for an additional mechanism of stranded cost recovery for IOUs through something called "intangible transition charges" (ITCs). On the first, the final law stated, "An electric distribution utility may seek, and the commission may approve, an exception [to the aforementioned rate caps]" (Pennsylvania House Bill 1509, §2804(4)(iii)) for a litany of reasons. These included being "unable to mitigate" certain stranded costs that had been expected to be mitigatable, increases in federal or state taxes imposed on the company, and other changes in wholesale market conditions. "Extraordinary rate relief" for any of these reasons could be provided through expedited proceedings which would likely limit opportunities for opposing parties to intervene (Pennsylvania House Bill 1509, §2804(4)(iv)). On the second, intangible transition charges could also be provided on an expedited basis, and at any time "concurrently with, prior to, during or following the filing of [the company's] restructuring plan," upon the request of an IOU for something called a "qualified rate order" to be issued by the PUC (Pennsylvania House Bill 1509, §2812).

These provisions amount to "off-ramps," "insurance policies," and other special treatment available only to IOUs and no other type of political actor. These instances of special treatment[23] fundamentally impact the distributive politics (Mildenberger 2020) of electricity policy, creating a situation in which IOUs almost always win, with or without the implementation of state-level restructuring. Recall the opening of chapter 1 in which Oregon IOU lobbyists recalled how their ability to *shape* that state's RPS legislation was fundamentally more important than whether it passed, or even than how ambitious its targets would be. The same can be said for IOU preferences over restructuring policies in states like Pennsylvania. As a former Pennsylvania PUC commissioner stated regarding IOUs' influence over the design of Pennsylvania's restructuring law: "Look, if you're in Harrisburg, walk over and look to see the registered lobbyists for utilities versus consumer advocates or versus industrial customers or commercial customers. I mean, utilities got a tremendous amount of money to lobby and got tremendous long-term relationships with members of the legislature over there. And they've got a lot of power" (author's interview, 5/15/23).

However, as we saw in the last chapter with the Texas case and as we will soon see again in the case of Pennsylvania, restructuring policies *do create* positive policy feedback effects (Stokes 2020) that impact IOUs' subsequent policy preferences when it comes to RPS. Specifically, state-level restructuring policies can create political conditions that are more favorable for RPS policies' success than they would otherwise be.

Getting HB 1509 through the legislature and signed by the governor required enormous discipline on the part of the stakeholders involved, and the key policymakers who led the negotiations played a role in enforcing such discipline. A former PUC commissioner recalled, "We had a meeting at the Governor's office, in the Governor's residence actually, and he went around the table, the governor did, and said, 'Alright. We've negotiated this package. Who's on board?' And, like, looked all the utility CEOs in the eye. And, you know, it was one of those great moments in gubernatorial leadership. . . . I mean, there were a few voices of dissention" (author's interview, 4/26/23).

A different former PUC commissioner recalled, of the final stage of the negotiations, "I said, 'Here are the rules. You can insist upon what you need, but you must be willing to negotiate on what you want. If you stand up from the table and leave the negotiations, you're out and can't come back. Anything that you get in the bill, if you leave, comes out [of the bill].' . . . And it was pretty contentious, as you can imagine" (author's interview, 5/15/23).

Importantly, another component of the "ground rules" that the stakeholders all agreed to was that floor amendments in the legislature would not be allowed to be adopted; substantive changes to the legislative text had to have been worked out privately before the bill was on the floor: "The rule was, Josh, everybody agreed to it . . . it's not amended by the legislature. Because, as you know, this stuff is so technical and so carefully balanced everywhere along the line. . . . So we went to the legislature and said to them, 'Look, we got a bill here, but it's highly technical. I only got agreement if it's not amended.' . . . And believe it or not, it was passed through both houses without amendment" (author's interview, 5/15/23). As noted earlier, final passage occurred in the wee hours of the morning and all significant amendments—including Senator Schwartz's, which would have established an RPS—were defeated on the floor (Pennsylvania Senate Journal 1996).

Once the bill was signed into law, the formal action moved to the PUC and the individual restructuring plans of the major IOUs. PECO and PP&L were the first to file, on April 1, 1997. Penn Power was the last to file, on September 19, 1997. The remaining companies all filed in the interim. Table 4.5 shows the number of pages for each of these filings as well as the number of pages of the resulting PUC Orders. Billions of dollars in stranded costs were recovered from Pennsylvania ratepayers through mechanisms like the CTC and ITCs (Pennsylvania Public Utility Commission 1997, 1997a, 1997b, 1997c, 1997d, 1997e, 1997f, 1997g, 1998, 1998a, 1998b, 1998c, 1998d, 1998e).

Table 4.5
Page counts for Pennsylvania restructuring filings of state's major IOUs

Company	# of Pages of Restructuring Plan Filing	# of Pages of PUC Order Approving Plan
PECO	4,333	267
PP&L	3,469	199
Duquesne	3,361	356
West Penn Power	2,642	307
UGI	3,164	45
Met-Ed	4,883	299
Penelec	3,378	307
Penn Power	282	75

Source: Pennsylvania Public Utility Commission.

Pennsylvania's Mandatory RPS

The 2003–2004 legislative session of the Pennsylvania General Assembly was significant in that it coincided with the start of Democratic Governor Ed Rendell's eight years in office. Republicans controlled both legislative chambers, but by very thin margins (see table 4.4). Rendell was previously the mayor of Philadelphia, a political stronghold for the state's Democratic Party, in which both organized labor and environmentalists are considered important constituencies. Education policy was the major issue Rendell had campaigned on, but upon being elected he had taken an interest in "alternative energy," and he appointed Katie McGinty, an executive branch "climate policy entrepreneur" (Rabe 2004) to lead his PDEP. By all accounts, McGinty was the main strategist in the executive branch who worked with several key, moderate Republican legislators to pass Act 213, Pennsylvania's "Alternative Energy Portfolio Standard (AEPS)" (author's interviews, multiple).

Before further discussing the political evolution and contents of Act 213, it is important to acknowledge and briefly discuss the name: the "Alternative Energy Portfolio Standard." Perhaps a rare instance of truth-in-advertising in modern-day politics, the title itself is an acknowledgment that, while mandatory, Pennsylvania's RPS was "dirtier" (Rabe and Mundo 2007, 279) than most. In particular, the act's name refers to its resource eligibility criteria, which are expansive and, as will soon be discussed, included certain technologies that are clearly *not* renewable—perhaps most obviously, "waste coal," a by-product of coal production. Glenna and Thomas (2010) provide a thorough and recommendable discussion of the politics of "policy labeling" when it came to the Pennsylvania Alternative Energy Portfolio Standard, a topic also discussed more generally by Rabe (2004). Suffice it to say, while it is not surprising that the political economy of Pennsylvania would render an environmentally *strong* RPS infeasible, it is still worth asking why Pennsylvania adopted a *mandatory* RPS at all, when other states in this book that produced considerably lower levels of fossil fuels declined to adopt one at all.

During the 2003–2004 legislative session in Pennsylvania, at least three bills to adopt an RPS were introduced: SB 962, HB 2250, and SB 1030. Only the first, SB 962, was introduced by a Democratic legislator; the other two were introduced by Republicans, who held a slim majority in both chambers at the time. SB 962, the Democratic bill, was the most environmentally stringent based on its targets/timeline and its narrower resource eligibility criteria (Pennsylvania Senate Bill 962), but it did not survive even its first committee in the state Senate. HB 2250 and SB 1030 were much more similar in terms of their content and language, as their two sponsors "worked cooperatively" on them (Pennsylvania House Journal 2004, p. 2256), and SB 1030 became the legislative vehicle that ultimately passed both chambers and was signed into law by Governor Rendell as Act 213 on November 30, 2004.

The state representative who was the chief sponsor and author of HB 2250, and who worked closely with the author of SB 1030, also a Republican, explained the importance of his Republican Party affiliation: "I was identified, I think, by the Rendell administration as a moderate Republican who was going to probably be helpful and willing and interested in working on renewable energy. . . . It was the classic situation of a Democratic governor with an environmental agenda dealing with Republican majorities in the House and the Senate, where, you know, a Democrat could have introduced a bill that was friendly to the administration and it would have died a discreet death" (author's interview, 5/2/23). When asked why *he* was willing to work with the Democratic administration, this state representative explained, "Oh, I liked the idea. I was very enthusiastic and still am about renewable energy, actually. . . . Before I got into the legislature I was pretty active with some environmental legislation and work, at the township level. So I was kind of tuned into some of this stuff already anyway. And I loved working on complex legislation" (author's interview, 5/2/23).

Clearly, we see the importance of partisanship in the careful selection of Republicans to steward the bill through the legislature. We also see the *possibility* of Republicans—at least back then, in the mid-2000s—championing renewable energy legislation in the state legislature. Admittedly, it was more precisely "alternative energy" legislation, but it did give an important boost to renewables, in ways that will soon be detailed. Moreover, it is evident from my database, as we will soon see, that it was Pennsylvania's *energy economy* rather than its partisanship that was responsible for the watered-down content of Act 213, which passed the House with seventy-five Republican votes and passed the Senate with fifteen Republican votes (Pennsylvania House Journal 2004; Pennsylvania Senate Journal 2004).

Act 213 was comprehensive in its coverage of IOUs, applying to all "electric distribution companies" as well as "electric generation suppliers." Its targets and timelines were modest relative to other states', yet still significant given Pennsylvania's abysmal starting point in terms of in-state renewable resources. By fifteen years after the effective date of the Act, 8 percent of electricity sold needed to come from eligible so-called tier 1 resources.[24] This target was broken down into smaller increments: 1.5 percent within two years; 2 percent within three years; and increasing by 0.5 percent annually every year thereafter. When it came to so-called tier 2 resources, 10 percent of electricity sold needed to come from these within the same fifteen-year timeline. Tier 2, also, was broken down into smaller increments: 4.2 percent by the end of year 4, 6.2 percent by the end of year 9, 8.2 percent by the end of year 14, and 10 percent by the end of year 15 (Pennsylvania Senate Bill 1030/Act 213, §3).

In addition, Act 213 included a very modest "solar carve-out." Half of 1 percent of the electricity sold by the end of year 15 needed to come from solar photovoltaic technology, a tier 1 resource, specifically (Pennsylvania Senate Bill 1030/Act 213, §3). While "tier 1" included resources that would be uncontroversial for a typical RPS (e.g., solar photovoltaic, wind, low-impact hydro, geothermal, and biomass), tier 2 was "loaded up with anything they could think of" (author's interview, 5/4/23). Specifically, this included large-scale hydropower, municipal solid waste, integrated combined coal gasification technology, and, most controversially, waste coal[25] (Pennsylvania Senate Bill 1030/Act 213, §2).

Regarding REC accounting, Act 213 did specify that RECs (in Pennsylvania, labeled "alternative energy credits") would be used as the compliance instrument, but it was silent on whether they needed to be bundled with the underlying electrons or whether they could be unbundled to increase compliance flexibility; that decision was left to the PUC (Pennsylvania Senate Bill 1030/Act 213, §3(e)). Act 213 established an ACP, which, except for the solar carve-out, was to be set at "$45 times the number of additional alternative energy credits [RECs] needed in order to comply" (Pennsylvania Senate Bill 1030/Act 213, §3(f)(3)). For the solar RECs, the ACP was set at "200% of the average market value" of the solar RECs (Pennsylvania Senate Bill 1030/Act 213, §3(f)(4)).

Additionally, as is typical across the RPS-adopting states (see also Basseches 2024), Act 213 contained cost/risk-shifting provisions to protect the regulated IOUs from financial uncertainty associated with these new mandates. These included a complete compliance exemption until the conclusion of the restructuring rate freeze and transition period (Pennsylvania Senate Bill 1030/Act 213, §2), the guaranteed ability to recover any additional compliance costs from ratepayers through an "automatic adjustment clause" (Pennsylvania Senate Bill 1030/Act 213, §3(a)(2)), and a "force majeure" provision, which allowed IOUs—"electric distribution companies," specifically—to petition the PUC to call the entire thing off under certain conditions that they argued could make compliance too onerous (Pennsylvania Senate Bill 1030/Act 213, §2). The political significance of these provisions will be discussed further in Chapter 5. Table 4.3 shows some key policy design elements of the Pennsylvania RPS.

In March of 2005, a few months after the passage of Act 213, the PUC convened an "Alternative Energy Portfolio Standards Working Group" of stakeholders to work out some of the finer details related to the implementation of the RPS program pursuant to Act 213. This resulted in the adoption of two "implementation orders," Implementation Order 1 and Implementation Order 2, on March 23, 2005, and July 14, 2005, respectively. Implementation order 1 included a customized compliance schedule for each IOU ("electric distribution company") that considered the end date of its rate freeze and transition period from its approved restructuring plan, since Act 213 stipulates compliance is required only beginning at that time. Implementation order 1 also developed rules for REC banking toward the end of IOUs' rate freeze periods, which were made even more generous (in terms of maximizing IOUs' flexibility) in Implementation Order 2. In Implementation Order 2, the PUC denied a request on the part of the IOUs, which competing stakeholders would have viewed as outrageous, to be able to recover in rates the cost of any ACPs, with the PUC writing, on page 13: "The alternative compliance payment appears to be intended to serve as a penalty provision that will encourage compliance with the Act [RPS, Act 213]" (Pennsylvania Public Utility Commission 2005, 2005a).

In 2007, with Rendell still governor but now also a House of Representatives with a slim Democratic majority for the first time since 1995, House Bill 1203 (which was signed into law as Act 35 on July 17, 2007) made only minor changes to the state's RPS. It broke down the solar carve-out timeline into smaller compliance increments—requiring annualized increases as opposed to increases every five years—but left unchanged the overall fifteen-year target (Pennsylvania House Bill 1203/Act 35, §2). In addition, House Bill 1203 clarified facility eligibility in terms of where the RECs could originate from geographically. They would be unbundled RECs that could originate from anywhere within the PJM Interconnection[26]; however, they could not be "double-counted," meaning that the same REC could not be used to also satisfy the RPS requirement of a different state (Pennsylvania House Bill 1203/Act 35, §4). In addition, House Bill 1203 clarified the requirements of "force majeure," instructing the PUC to apply a "good faith effort" standard in determining whether an IOU should be relieved of its RPS compliance obligations (Pennsylvania House Bill 1203/Act 35, §2). This will be discussed further in chapter 5.

The PUC issued a single implementation order regarding Act 35, on September 25, 2008. This order rejected a request from environmental groups and independent power producers that the rules around the "force majeure" provision be constructed more narrowly. In a win for IOUs, which will be further discussed in the next chapter, the PUC declined this request (Pennsylvania Public Utility Commission 2008). In the 2009–2010 legislative session and beyond, several bills appeared in Pennsylvania to increase the RPS, including increasing the solar carve-out. These included HB 80, SB 92, and HB 2405; none made it through even its legislative chamber of origin, despite consistent support from environmental groups and the small but growing renewable energy industry within the state.

HB 80 had the most political momentum of the three. It would have increased the amount of tier 1 ("cleaner") electricity required beginning in

2014, ending in 2024 at 18 percent. It would have also *significantly* increased the solar carve-out, beginning in 2011 and ramping up to 3 percent by 2024. It would have also limited facility eligibility for solar to within the state's borders (rather than anywhere in the PJM region), encouraging the development of new solar generation *in Pennsylvania* rather than in adjacent states with their own, more ambitious RPS policies already in place. In short, HB 80 would have constituted a major strengthening of Pennsylvania's mandatory RPS (Pennsylvania State House of Representatives Archives 2009a).

Although HB 80 was strongly supported by environmental groups and "green business" actors, that support was overwhelmed by opposition from "traditional business" actors (Basseches 2024). These traditional business and fossil fuel industry actors' interests were represented by Republican legislators who, by this point post–Tea Party rise, had become more staunchly opposed to climate and renewable energy policy (Pennsylvania State House of Representatives Archives 2009a, 2009b, 2010).

Pennsylvania Within-Case Analysis

Before moving on to the comparative analysis of Oklahoma and Pennsylvania, which will focus on the role of a restructured electric utility sector in Pennsylvania but not Oklahoma, a within-case analysis of Pennsylvania is in order. Here, there are two noteworthy dynamics. First, the environmental advocacy community in Pennsylvania was divided over SB 1030 because some groups took the position that *any* mandatory standard was preferable to none, even if that mandatory standard included questionable resources such as waste coal, while others opted to oppose the bill outright so long as it included nonrenewable resources. Second, despite Pennsylvania ostensibly being a "red state," it was the state's energy economy more than its partisan composition that was most responsible for limiting the strength of its mandatory RPS policy.

At the very beginning of the 2003–2004 legislative session, a unified environmental advocacy community welcomed the RPS legislation. For instance, the Clean Air Council, which would later oppose the legislation once nonrenewable resources were included, wrote to the House Energy and Environmental Resources Committee on April 13, 2004: "The creation of a Renewable Portfolio Standard is a matter of great importance to the economic welfare of the Commonwealth [of Pennsylvania] and the health of its residents. . . . Clean Air Council is pleased to see the serious consideration renewable energy is receiving in the state Legislature. HB 2250 is a good bill upon which to build" (Pennsylvania House of Representatives Archives 2004).

However, as the legislation picked up momentum as the session continued, the fossil fuel industry and "traditional business" actors (Basseches 2024) were mounting increasing pressure to kill it. The waste coal industry in the form of ARIPPA saw an opportunity to secure a future for itself while watering down the legislation from an environmental perspective. On November 15, 2004, ARIPPA wrote to the Senate Appropriations Committee, the Senate Environmental Resources and Energy Committee, and the House Environmental Resources and Energy Committee: "We express our support for including coal refuse [waste coal] in the Alternative Energy Portfolio Standard ['AEPS'] currently under discussion. . . . If the AEPS is properly crafted, including waste coal as an eligible resource will help ensure that existing waste coal plants will be able to continue to operate as they are forced to compete with large conventional units in the open market. It also will provide a needed incentive to accommodate the construction of new waste coal plants as Pennsylvania's demand for power increases" (Pennsylvania House of Representatives Archives 2004).

Meanwhile, state policymakers saw an opportunity, by including waste coal as part of tier 2, to remediate piles of abandoned ash that were threatening to pollute rivers and other natural resources. As a lobbyist who at the time was in close contact with key legislators and members of the Rendell Administration negotiating Act 213 recalled: "You used to drive past these mountains of waste coal in Pennsylvania. You know, if you went up [Interstate] 81, there's a little turn on there, there's a little place called Ravine, and a little further down the road on the left was this huge waste [coal pile]. . . . That's [now] gone. So, the argument was that it was performing an environmental benefit by getting this stuff off the landscape and using it to generate electricity in boilers" (author's interview, 4/13/23).

Several environmental groups did not see it that way; they saw it instead as the squandering of an opportunity for Pennsylvania to become a leader in the promotion of truly *renewable* energy, on par with what neighboring states like New Jersey had already enacted. As the leading advocate from one of these groups explained:

> In all honesty, while I tend to think on most environmental policies the environmental community stays together, on this particular issue, there was a fair amount of disagreement on the policy that was moving, you know, the AEPS Bill. . . . For many of us, we thought the bill was antiquated before the ink dried when Governor Rendell signed it, and it set some really bad precedent. You know, it's a two-tiered system. . . . A top tier at 8%, and the dirty tier two, which was the first time, I believe, any state portfolio [standard] had done that. The dirty part [tier 2] was actually larger. . . . So I think for many of the groups, PennEnvironment, Sierra Club, Clean Water Action, Clean Air Council, I think many of us became fairly disillusioned with feeling that the negative pieces outweighed the positives." (author's interview, 4/26/23)

Policymakers who were aware that the strength of the fossil fuel industry in the state would inevitably limit the policy's ambition were still intent on getting *something* that was mandatory passed, and they were frustrated that so many environmental groups were unwilling to lend any support, once tier 2 had been negotiated. One of them, who made no secret of his frustration even all these years later, stated:

> Some of them [environmental groups not supporting the 2004 RPS bill] have a very hard time getting to "yes." . . . Let's talk about Tier 2 for a moment. It's 10%. The moment that law was signed in 2004, the supply into Tier 2 exceeded the 10% requirement. . . . And we knew this was absolutely giving away, at least in that time period, nothing. And you could explain that all you wanted and show the math to PennEnvironment, which was one of the organizations that opposed it, and it wouldn't matter. (author's interview, 5/5/23)

The state representative who sponsored HB 2250 (which became SB 1030) recalled: "The Sierra Club never endorsed our bill. . . . Because the Sierra Club, they were really p'd [pissed] off about the waste coal piece and other environmental groups were also. . . . And so, that was a problem, you know, over on the Democratic side. And so you weren't really being thanked for bringing this forward as much as you might have been by the environmentalists. Now, mainstream environmental groups did eventually come on board and help us" (author's interview, 5/2/23). One such so-called mainstream environmental group was PennFuture. Rendell's point person on the legislation described that group as "sophisticated" about understanding the political reality and being skilled at the art of legislative negotiation and compromise (author's interview, 5/23/23).

Still, a divided environmental community was a key feature of the politics of the Pennsylvania AEPS. And such internal disagreement has continued through more recent attempts to enhance the RPS (which, unlike SB 1030, were not ultimately successful). For example, when it came to HB 80 in the 2009–2010 legislative session, the Sierra Club, PennEnvironment, Clean Air Council, and Clean Water Action all conditioned their support of that bill on resource eligibility requirements, including the exclusion of nuclear energy and carbon capture and sequestration (CCS). By contrast, groups like the Pennsylvania Environmental Council and PennFuture offered support that was less conditional (Pennsylvania State House of Representatives Archives 2009, 2009c).

In understanding the limits of Pennsylvania's RPS/AEPS, beyond the divided environmental community, the data shows that the state's energy economy and the interest group politics that resulted from it mattered more than Pennsylvania's ostensibly Republican-dominated partisan politics. Indeed, were it not for the political strength of the fossil fuel industry and "traditional business" actors (Basseches 2024), the concept of tier 2 would likely have never even been on the table in the first place. While it was not a coincidence that Act 213 passed in the first session that a new, Democratic governor was in office, and that the next time the legislature modified it was during the first session in quite some time that Democrats controlled the House, in 2007, the evidence suggests that the importance of partisanship was dwarfed by the importance of sustained lobbying on the part of the fossil fuel industry and other "traditional business" actors. The archival record, along with the totality of the interview data, paints the picture of a powerful interest group community opposed to renewable energy mandates throughout the entire period under study. The following testimony from the Pennsylvania Coal Association on HB 80 summarizes its position aptly: "In recent years [Pennsylvania's] government has, through regulations, subsidies and mandates, again forced its influence on the market in a manner that skewed the playing field against the use of coal. Among the intrusions into the customer choice market was the 2004 enactment of Pennsylvania's AEPS" (Pennsylvania State House of Representatives Archives 2009c).

It wasn't just the Coal Association that fiercely opposed a stringent RPS in Pennsylvania; it was also other "traditional business" actors (Basseches 2024), which, in Pennsylvania's case, included the Chamber of Business and Industry, the Industrial Energy Consumers of Pennsylvania, the Pennsylvania Manufacturers Association, the Associated Petroleum Industries of

Pennsylvania, the National Federation of Independent Business, the Pennsylvania Electric Power Generation Association, the Pennsylvania Chemical Industry Council, the Pennsylvania Oil and Gas Association, and the Energy Association of Pennsylvania. All of these groups cosigned a letter to the chairman of the House Environmental Resources and Energy Committee, which read, in part, "On behalf of our thousands of Pennsylvania employers, from Sole Proprietors to Fortune 100 companies, the undersigned write in opposition to HB 80, amending the Alternative Energy Portfolio Standards Act (AEPS Act) . . . Pennsylvania should steer clear of government mandated energy requirements" (Pennsylvania State House of Representatives Archives 2009d).

As one lobbyist who was involved at the time put it, "All the big guys [opposed a stringent RPS]. All of the big guys" (author's interview, 4/13/23). As a former state policymaker put it: "The real dominant reason [that the RPS was not more stringent] is that Pennsylvania is a [fossil fuel] energy-producing powerhouse" (author's interview, 5/5/23). Even in comparison to the industrial customers' association (which also opposed a stringent RPS), which across the seven states in this book is a very influential actor, this policymaker said of its political influence relative to that of the coal and gas industries: "It's like comparing the eight men on an NBA roster to a team's all-star. . . . They [the industrial customers' association] were a player. . . . They certainly have more power than any single environmental group or probably collection of environmental groups in the legislature or at the PUC. I don't want to discount them, but I'll stick with my eight-man on the NBA—they play the game, they get off the bench. The coach calls them, but they're not deciding the games" (author's interview, 5/5/23).

Moreover, it is clear from interviews with key negotiators of the text of Act 213 that the fossil fuel industry, led by the Coal Association, was directly responsible for moderating the ultimate law's ambition. As one of them recalled: "The coal industry had successfully whittled down the 15% Tier 1, which was the renewable tier, to 8%. They used their political clout to get it down to 8%" (author's interview, 5/5/23).

A different policymaker recalled, in the following exchange:

> **Interviewee:** "The real key thing was relationships. . . . And so I had joyously spent a lot of time of course, you know, with the environmental groups with whom I, you know, I'd work forever and ever seamlessly. But also with the, as we fondly called them, 'the coal guys.'"

Interviewer: "So is that where the Tier 1 and Tier 2 construct came from?"

Interviewee: "Well, once the relationships of trust, you know, started to be there and then this other backdrop of momentum. I mean, this was a multiply orchestrated thing, but the economic opportunity, and getting those coal guys to know that I was out to build a positive and not to paint them as some demonic negative, you know. . . . So there was many a beer and many a Johnny Cash song listened to and enjoyed with those guys who, some of whom are still very good friends. But once that relationship with trust was built, then it created the room to get imaginative about a legislative structure. . . . The fact that, you know, the coal guys were never going to cheer for it [the bill], but you know, they essentially found a way to get themselves out of town."

Interviewer: "They stood down?"

Interviewee: "They did. They did." (author's interview, 5/23/23)

A lobbyist representing one of the interest groups that cosigned the aforementioned letter opposing HB 80 confirmed that his client's interests were, to a degree, accommodated: "We didn't think it [an RPS/AEPS] was necessary at all. What we did not oppose was—in fact, a lot of the people who pushed to get the Tier 2 segment in there were our members. . . . If you keep it at just wind and solar our view is that it gets very expensive. If you say, well, you can satisfy your requirements by doing this, you know, by adding any of these five things, in our view, that will tend to work to mitigate those expenses" (author's interview, 5/25/23).

At the same time, the evidence suggests that partisanship did *not* play a decisive role in the ultimate design of the Pennsylvania RPS. Not only was Pennsylvania less of a "red state" and more of a "purple state" relative to others in this book during the period under study, and not only was Act 213 sponsored by self-described "moderate Republicans," but also, the Republican Party in Pennsylvania simply did not openly resist renewable energy until approximately the 2010 election, long after Act 213 had been implemented. The following quotes, from three different policy advocates and lobbyists describe well the relationship between partisan politics in Pennsylvania and renewable energy policy in the pre- and post-2010 periods:

"It was the 2010 gubernatorial election when Republicans first ran against renewable energy." (author's interview, 5/16/23)

"Frankly, it was the Tea Party movement and the election of a crop of extreme, you know, kind of right-wing folk into the legislature [in 2010] that began to develop this stupidity that we have now." (author's interview, 4/26/23)

> "Back then, you know, moderate Republicans could be reached because there *were* moderate Republicans. And so there was sort of a formula, you know you needed [to get renewable energy policy done], and you always knew who they were. So you needed to peel these people [moderate Republicans] off, and hold as many of the Democrats as possible, knowing you'd lose some of the Western guys [Democrats loyal to coal]. So that's what you had to do." (author's interview, 4/13/23)

Oklahoma Versus Pennsylvania: The Pivotal Role of Utility Sector Structure

Given the similarly "fossil-fueled" economies of Oklahoma and Pennsylvania, and the interest group politics they created—with coal/oil/gas industries being politically powerful and staunchly opposed to a stringent RPS in both states—how can we understand why Pennsylvania ended up with a mandatory RPS policy while Oklahoma ended up with only a voluntary one? As has already been shown, unlike in Oregon and South Carolina—in which partisanship played the most significant role in explaining those states' divergent outcomes—in neither Oklahoma nor Pennsylvania was partisanship a particularly salient factor. While Pennsylvania certainly had a stronger presence of environmental groups than did Oklahoma, as has been shown, they were not particularly forceful when it came to the Pennsylvania RPS specifically, since they were divided over the resource eligibility requirements.

Instead, in the remainder of this chapter, I will argue that the best way to understand the divergent outcomes in Pennsylvania and Oklahoma is in terms of the policy feedback effects (Pierson 1993) of Pennsylvania's decision to restructure its electric utility sector and Oklahoma's decision not to. While Oklahoma's two vertically integrated IOUs were strongly opposed to any renewable energy mandate and were extremely politically influential, in Pennsylvania, IOUs' preferences were more divided, and their exit from the generation business created a dynamic—similar to the one in Texas—in which they were less able to provide a powerful, unified force in opposition to a mandatory RPS.

Describing an earlier attempt (which failed) to advance a mandatory RPS bill through the Oklahoma legislature, a wind industry lobbyist recalled:

> The RPS bill that we worked on. . . . It had to pass through the Utilities Committee in the House, and then it would go to the [House] floor. So, we got on the agenda for a meeting of the Utility Committee. It was mid-session, early in the

> session. And we did a nose count. We'd done enough lobbying that we had about 60% of the committee that agreed that they would vote for it. The utilities found out about that, talked to the Chairman who moved the meeting up by half an hour. . . . He moved the meeting up half an hour, only let the people who were gonna vote against it know that the meeting had been moved up half an hour, put it first on the agenda, voted it down. That was like, God damn." (author's interview, 2/17/23)

A top legislative staffer who worked on HB 3028, the purely voluntary RPS bill that eventually became law in 2010, recalled specifically that Oklahoma's two major vertically integrated IOUs were both actively involved as the legislation was being crafted, and both had a strong interest in maintaining language that would render the RPS purely voluntary. He recalled:

> Both PSO, Public Service Company of Oklahoma, and OG&E [Oklahoma Gas and Electric] were involved. . . . in our discussions. As I recall, they had initial concerns with some of the initial wording. . . . And [they] didn't want to mandate. . . . I do remember sitting across the table with the Speaker [of the House] and a guy named "Mike"[27] at OG&E at that time. He was their lobbyist . . . But "Mike," you know—I remember we were very careful about every word we used. (author's interview, 2/27/23)

A former PSO lobbyist recalls his company was similarly opposed to an RPS *mandate*: "We did oppose [a mandate] at the time. And we set a goal, but not a standard" (author's interview, 3/17/23).

In Oklahoma, like in South Carolina and Oregon but unlike in Pennsylvania and Texas, monopoly IOUs owned significant fossil-fueled generation assets. In other work, I found that this was an important factor in shaping their subsequent policy preferences when it came to climate and renewable energy proposals (Basseches 2024). Large-scale quantitative analysis supports this proposition as well (Hall et al. 2024).

It is clear from the data that, in the Oklahoma case, the traditionally regulated, vertically integrated system resulted in IOUs' opposition to a mandatory RPS. Of course, in the Oklahoma case—quite different from the South Carolina or Oregon cases—the IOUs were hardly alone in their opposition to a mandatory RPS; the enormous oil and gas industry presence in Oklahoma was another factor that took a mandatory RPS off the table, and it is impossible to say what the outcome would have been in Oklahoma if *only* the IOUs were against it and not the oil and gas industry as well. Nevertheless, what is important here in differentiating the Oklahoma case from the Pennsylvania case is that Oklahoma's major IOUs were vertically integrated and unified in their opposition to a mandate. As "800-pound gorillas in the legislature" (author's interview, 2/17/23), according to one lobbyist who described the two major IOU companies this way, their preferences carried outsized weight when it came to policy design.

By contrast, in Pennsylvania, the story was quite different. When Act 213 was under consideration during the 2003–2004 legislative session, Pennsylvania's IOUs were all in the process of restructuring pursuant to the 1996 Electricity Generation Customer Choice and Competition Act and had either sold off their generation assets completely or were in the process of doing so. They were also in the process of recovering their stranded costs from ratepayers pursuant to PUC approval of their individual restructuring plans. As a result, their policy preferences regarding a renewable/alternative energy mandate were more mixed, but overall neutral. This is not to say that they all *welcomed* the act, but, on the whole, Pennsylvania IOUs perceived it to be far more innocuous compared to their counterparts in Oklahoma. As a consumer advocate who intervened regularly at the Pennsylvania PUC put it, characterizing the state's IOUs overall rather than any one particular company: "I think that they [IOUs] were skeptical, I guess, of the RPS or AEPS, but they weren't you know—they probably couldn't have passed it if they had been adamantly opposed to it" (author's interview, 4/7/23).

When it came to particular Pennsylvania IOUs, their positions on Act 213 varied from positive to neutral and/or cautious. On the more positive end, a Duquesne Power and Light spokesperson stated: "Duquesne Light has been and continues to be a proponent of renewable energy portfolio standards. *We're unbiased supporters because we have no generation* [emphasis added]" (Stouffer 2004). A spokesperson for Allegheny Energy, parent company of West Penn Power before it was acquired by FirstEnergy in 2010, stated: "We worked in cooperation with the General Assembly to put together a fair law, and we will comply with it" (Stouffer 2004).

PECO, in legislative testimony co-signed by its parent company Exelon, expressed very cautious support for an RPS in Pennsylvania, warning that it might raise electricity costs for the company's *customers*, but suggesting it would not pose any threat to its shareholders, since plans were already in motion for PECO's exit from the generation business. That testimony read, in part: "Exelon and PECO support renewable energy resources *in a competitive electricity market* in Pennsylvania. . . . In recent years several states have adopted RPS programs and Pennsylvania is in the early stages of looking at RPS. As we begin to evaluate both the pros and cons, as well as the process by which economic renewable energy generation enters the marketplace, the following issues need to be addressed . . . [emphasis added]" (Pennsylvania State House of Representatives Archives 2004).

The letter then goes on to enumerate various concerns having to do with the possibility that customers' electric bills may go up as a result, that the company be allowed to recover any above-market costs from ratepayers, and so forth. Importantly, note the language in the letter; it suggests the company's position is contingent on the "competitive electricity market" enabled by the state's 1996 restructuring law; had the state remained traditionally regulated with vertically integrated utilities, the position that companies like PECO may have taken may have been more in line with what we saw in Oklahoma. In the words of one lobbyist, "Vertical integration was a problem [in terms of PECO supporting renewable energy mandates]," but "when that ended and PECO became just a wires company, technically agnostic to the source of the supply, then they softened their opposition" (author's interview, 5/5/23).

At the same time, the interest group politics in Pennsylvania were somewhat complicated by the fact that certain IOUs (electric distribution companies) were owned by energy *parent* companies that *also* had an interest in the generation business through various subsidiaries. Depending on the generation subsidiary and the degree to which it was invested in fossil-fueled assets, the parent company's position toward a mandatory RPS may have been more hostile than the associated distribution IOU's. This would explain why the Electric Power Generation Association, whose member companies included Exelon Generation,[28] FirstEnergy Corporation,[29] and PPL Generation,[30] among other nonaffiliated companies, offered much more nuanced and generally less supportive testimony on SB 1030 (Pennsylvania State House of Representatives Archives 2004). An interview with a lobbyist for a utility trade association that represented *both* distribution IOUs *and* their parent companies (some of whom also owned generation affiliates) confirmed how this complicated the politics: "Your perspective as a standalone distribution utility without any affiliated generation is different than when you're part of a family of companies" (author's interview, 5/15/23).

Nevertheless, clearly in contrast to Oklahoma, in Pennsylvania, the IOU industry was not uniformly opposed to a mandatory RPS. Indeed, in some cases, key segments of it were supportive. Just like we saw in the previous chapter in the case of Texas, this fragmentation of the IOU industry in Pennsylvania also yielded a new landscape of interest group power politics, consistent with policy feedbacks theory (Pierson 1993), in which even those forces within the IOU industry that were less enthusiastic about an RPS mandate constituted a less politically powerful bloc than they would have been had the industry not been restructured.

An interview with a former PUC regulator illustrates well, in the case of PECO, how its political power was very different before restructuring compared to after. Asked whether PECO was influential within the legislature before restructuring, he replied:

> Hugely. Very doesn't even begin to describe it. Hugely. They had spent a lot of money directly in politics for decades, but they had also, obviously, provided lots of jobs, including as favors to various politically important people, legislators and others, their families, their supporters. It was essentially a form of patronage. If you were on the side of PECO and you were a legislator and you had a constituent who needed a job, one of the places you would end up calling probably was the Philadelphia Electric Company. (author's interview, 4/28/23)

However, in the aftermath of restructuring, the story was different: "Even by 2004 [when Act 213 passed], PECO was a very different animal. Restructuring had turned PECO from an 800-pound gorilla in Pennsylvania politics to—it cut its political power by about 80%. It still mattered, but it was nothing like it was before" (author's interview, 4/28/23).

Speaking about the IOUs more generally, a former representative of industrial customers, who frequently butted heads with the utilities on a range of policy issues, stated: "[The political power of the utilities] substantially declined [after restructuring], yeah. Because markets are dictating, now. . . . And you know, when you think of the resources that the utilities are now dedicating at the state level, it pales in comparison to what it did pre-restructuring. You know, sure, they're still active politically, but if you would go back and look at their political contributions to the legislators and gubernatorial races in the '90s versus today, I'd bet it's half or less" (author's interview, 5/24/23).

And so, in this chapter we have seen—through an in-depth examination of the Oklahoma and Pennsylvania cases—the importance of electric utility sector restructuring relative to energy economy and partisanship when it came to RPS policy design. While partisanship played a much smaller role in both cases than it did in explaining differences between Oregon and South Carolina in the previous chapter, and energy economy played a much larger role in both cases, the most important factor in explaining why the "coal-and-gas state" of Pennsylvania adopted a mandatory RPS policy while the "oil-and-gas state" of Oklahoma adopted only a voluntary one was that Pennsylvania restructured its electric utility sector while Oklahoma did not.

Although it is true that Pennsylvania's RPS was environmental weak relative to most other states' given its resource eligibility requirements (Glenna and Thomas 2010; Rabe and Mundo 2007), it is also true that it was nevertheless a *mandatory* standard, which puts Pennsylvania in a unique position relative to other states with equally unhospitable political environments (such as Oklahoma). Although Pennsylvania's RPS policy was ultimately less impactful than Texas's (Langniss and Wiser 2003), it still moved the needle. As one renewable energy lobbyist put it: "[Act 213] was significant in terms of putting Pennsylvania on the map early on. . . . It was huge in the sense that it got attention on a coal state, a big coal state. . . . So what happened there was the AEPS got attention on Pennsylvania, and then the money [from the renewable energy industry] came after" (author's interview, 5/16/23).

In this chapter and in the one before it, we have seen how IOUs' policy preferences on renewable energy policy can vary across states, depending in part on utility sector structure. We have also seen that fully restructured utility sectors reduce the political power of those segments of the IOU industry that oppose renewable energy mandates. Nevertheless, this should not necessarily be interpreted as meaning that IOUs overall are less influential, even in fully restructured states, nor that their preferences vary when it comes to the *distributive* quality of RPS policy design. In previous work, I suggest it is useful to consider climate and renewable energy policy design on two dimensions: environmental strength and who pays (Basseches 2024). Until now, this book has primarily emphasized the first dimension, environmental strength. In the next chapter, we turn to the second dimension—who pays—and find that IOUs never pay; it is always ratepayers that pay, whether or not the state has restructured.

5 How IOUs Always Win on Distributive Politics

Over the previous two chapters we have seen how partisanship and energy economy each play a role in shaping prospects for state-level RPS policy design. However, at the same time, we have also seen how the structure of the electric utility sector is a pivotal and underappreciated determinant as well. Utility sector structure matters because it can determine whether IOUs are unified in opposing renewable energy mandates, or alternatively, whether their policy preferences are divided according to the niche within a restructured environment that a particular IOU occupies, as we saw was the case in Texas and Pennsylvania.

In this chapter, we focus more squarely on IOUs, which are pivotal actors with enormous power to shape state-level policy, irrespective of a given state's utility sector structure. While their policy preferences regarding the environmental strength and ambition of the RPS may indeed vary more in restructured states than in vertically integrated ones, when it comes to distributive politics—that is, the economic winners and losers associated with policy design choices (Mildenberger 2020)—the IOUs are always winners. Indeed, in this chapter, we will see how—even in the states that *did* restructure—restructuring policies were largely negotiated on incumbent IOUs' terms. In states that subsequently adopted mandatory RPS policies, IOU firms that supported these policies not only supported them but *shaped* them, shifting any potential costs or risks that such policies might confer from their own shareholders onto ratepayers (Basseches 2024).

In previous work, I argued that in evaluating the quality of state-level climate and renewable energy policies, it is important to analyze the policy not only in terms of its environmental stringency (i.e., its likelihood to reduce GHG emissions) but also in terms of its distributive quality—that is, who pays—or who assumes financial risk—for even the most environmentally stringent policies (Basseches 2024)? Certainly, in the medium to long run, *society as a whole* stands to realize enormous economic savings from climate policy action, not only with the costs of clean energy rapidly falling but also in terms of the avoided costs of unmitigated climate damages (e.g., insurance costs, natural disaster recovery, resource scarcity, etc.). However, as we know by now, politicians and policymakers rarely think in terms of the medium to long run, nor do they often think in terms of costs and benefits to *society as a whole*. Rather, they tend to think in terms of the short run (i.e.. what will allow them to maintain their positions of power through the next election cycle?) and they tend to think in terms of costs and benefits to *particular, organized political constituencies* (i.e., interest groups) with the ability to help them or obstruct them in their pursuit of this goal (Bawn et al. 2012; Cullenward and Victor 2020; Mildenberger 2020; Stokes 2020). In other words, politicians and policymakers are primarily concerned about distributive politics (Aklin and Mildenberger 2020; Wilson 1980).

In this chapter, I argue that when it comes to electricity sector policies such as RPS, IOUs are unmatched in their ability to win on the distributive politics ("who pays?") dimension of policy design (Basseches 2024). I show how this is true regardless of utility sector structure, partisanship, or energy economy. In traditionally regulated, vertically integrated states, IOUs tend to have an easier time keeping a mandatory RPS off the agenda altogether (although, as the Oregon case shows, and as other cases not examined in this book[1] show, utility sector structure is hardly the *only* factor that determines this).

However, regardless of utility sector structure and regardless of whether IOUs succeed in keeping an RPS off the agenda, IOUs tend to succeed in redistributing costs/risks associated with RPS policies away from themselves and onto other actors—often, ratepayers. I argue that their consistent ability to do so stems from a combination of structural and instrumentalist forms of political power distinctive to their industry, which I elaborate in chapter 6.

In the current chapter, I begin by illustrating the political power of vertically integrated IOUs in the two states in this study in which IOUs succeeded in keeping a mandatory RPS off the agenda entirely: South Carolina and Oklahoma. I then illustrate their power in Oregon, where even though they were unable to keep a mandatory RPS off the agenda due to that state's

Democratic politics and the lack of an in-state fossil fuel industry (see chapter 3), they still managed to shape the policy's design in a way that offloaded any costs/risks to their shareholders onto other actors. Recall the vignettes from Oregon which this book opened with, and the "five basic needs, must haves" (author's interview, 2/26/18), which IOU lobbyists secured in the initial RPS law in 2007.

I then examine IOUs' influence over RPS policy design in four restructured states: Texas, Pennsylvania, California, and Massachusetts. I focus especially on Texas and Pennsylvania, since California and Massachusetts are discussed elsewhere (see Basseches 2024). In all four of these states, IOUs shaped the distributive politics of *both* the initial restructuring policies *and* the subsequent mandatory RPS policies, which at least some segments of the IOU industry supported in all cases due to the feedback effects of the prior restructuring policies (Pierson 1993). Although the restructuring policies in these states had rendered at least some IOUs agnostic to the environmental strength of the mandatory RPS, the same was *not* true when it came to the question of "who pays?" (Basseches 2024). On *this* question, all IOUs were highly mobilized and highly effective in achieving their policy preferences in law and regulation.

South Carolina and Oklahoma: No (Mandatory) RPS, No Problem

South Carolina

As we saw in chapter 3, South Carolina never adopted a mandatory RPS, but that did not stop the state's powerful, vertically integrated IOUs, Duke and Dominion,[2] from offloading clean electricity policy costs and risks onto their ratepayers. We will look at two ways in which they achieved this: first, through the passage of the 2007 Base Load Review Act, and second, through the cost recovery provisions of the voluntary RPS legislation in 2014.

In the early to mid-2000s, South Carolina's IOUs (along with IOUs in other southeastern US states) were interested in building new nuclear plants to provide "base load" electricity[3] in the face of upcoming power plant retirements and increasing demand forecasts. Construction of new nuclear plants is extremely expensive; this is good for IOUs, with their business model allowing them to earn a regulated return on investment as a percentage of their capital costs (including construction of new power plants), but not so good for the ratepayers, who end up paying these costs, plus that percentage return. A South Carolina IOU lobbyist explained the conundrum IOUs faced in fronting such high nuclear construction costs:

> As a capital-intensive business, the financing you get matters a whole lot, and the terms for your financing. And you're more likely to get better terms for financing if you can show your lenders, or your debtholders, that you're able to start to get recovery [from ratepayers] on the asset before it goes into service . . . And getting paid for an asset that's not yet in service is novel, or was novel [at the time]. And they called that "construction work in progress," CWIP. . . . And we felt that was necessary in order to begin our nuclear project. (author's interview, 5/22/22)

This "problem" of financing was solved when the South Carolina state legislature enacted SB 431 (Act 16), the Base Load Review Act, which took effect May 3, 2007. This law not only granted cost recovery for CWIP but also contained extremely IOU-friendly language throughout. For example, §225 of article 4, which the new law added to chapter 33, title 58 of the 1976 Code states:

> (E) Unless the record in a subsequent [public service commission] proceeding shows that individual items of cost were imprudently incurred, or that other decisions subsequent to the issuance of a project development order were imprudently made considering the information available to the utility at the time they were made, *then all the pre-construction costs incurred for the potential nuclear plant must be properly included in the utility's plant-in-service and must be recoverable fully through rates* in future proceedings under this chapter [emphasis added]. (South Carolina Senate Bill 431/Act 16 2007, §2)

The law then goes on to provide for an "insurance policy" of sorts, should the utility decide to abandon a project after it is deemed imprudent:

> (G) If the utility decides to abandon the project after issuance of a prudency determination under this section, then the pre-construction costs related to that project may be deferred . . . [and] recovery of capital costs and the utility's cost of capital associated with them may be disallowed *only to the extent* that the failure by the utility to anticipate or avoid the allegedly imprudent costs, or to minimize the magnitude of the costs, was imprudent considering the information available at the time that the utility could have acted to avoid or minimize the costs. Pending an order in the general rate proceeding or revised rates proceeding, the utility, *at its discretion*, may commence to amortize to cost of service the balance of the pre-construction costs related to the abandoned project over a period equal to the period during which the costs were incurred, or five years, whichever is greater [emphasis added]. (South Carolina Senate Bill 431/Act 16 2007, §2)

Then, in §290, the Base Load Review Act offers regulated IOUs further agency to continue ahead with lucrative (to them, but expensive to ratepayers) construction projects even in the absence, or pending the reconsideration of, a so-called "project development application":

> The denial of a project development application . . . under this article shall not preclude the utility from filing a new or amended project development application . . . at any time. *A utility may proceed to construct a plant even if assurance of prudency or cost recovery is not sought or is denied, and the failure to seek or obtain such assurance may not be used as evidence or precedent in any future proceeding* [emphasis added]. (South Carolina Senate Bill 431/Act 16 2007, §2)

It was cost/risk-shifting language like this, which appeared in the Base Load Review Act of 2007 and which could not have plausibly benefitted any interest group actor other than the vertically integrated IOUs themselves, which enabled subsequent scandals such as the VC Summer debacle of 2017, in which ratepayers paid more than $2 billion for a nuclear plant which never produced any electricity (Kinnard and Collins 2018).

Then, in 2014, when South Carolina's voluntary RPS was adopted in the form of SB 1189/Act 236, not only was it relatively toothless from an environmental standpoint, but it was also a risk-free proposition for the state's IOUs, which is why they did not lobby against its passage. The legislation was preceded, as is often the case, by an informal stakeholder negotiation process, in this case known as the Energy Advisory Council (EAC). An IOU lobbyist, "Miranda,"[4] recalls her charge going into the meetings of this EAC: "My involvement was just, you know, they [managers at the company] said, 'Miranda,' go figure out what this EAC is about and make sure nothing bad happens. . . . Miranda, go down to Columbia[5] and attend these EAC meetings and tell me what's going on.' First order of business is, like, don't let bad stuff happen. And second order of business is, figure out if there's any good stuff that can come out of it" (author's interview, 5/22/22).

"Miranda" executed her mission impeccably. Not only did the resultant legislation contain no RPS *mandate*, but even the voluntary programs that IOUs could opt into to promote renewables came with legislative language that made explicit that there would be no cost, or no "lost revenue," associated with their decision to do so. Section 2 of SB 1189 states: "An electrical utility shall be permitted to recover its costs related to the approved distributed energy resource program" (South Carolina Senate Bill 1189/Act 236 2014, §2). But not only this, even if an IOU were to opt to *terminate* a voluntary program, the company would *still* be entitled to full cost recovery for any costs up to the point of termination: "The effect of a decision to amend or terminate an approved distributed energy resource program, investment, or measure *shall be prospective only, and costs incurred prior to that decision shall be recoverable* [emphasis added]" (South Carolina Senate Bill 1189/Act 236 2014, §2).

Similarly, if the *state* were to terminate (by not renewing) the voluntary RPS program, any associated costs would *still* be fully recoverable by the participating IOU: "(F) The authorization to propose or approve new components of DER programs shall sunset and expire on January 1, 2021, *provided however that the cost recovery provisions of this chapter shall remain in force until the costs associated with all DER program components have been recovered* [emphasis added]" (South Carolina Senate Bill 1189/Act 236 2014, §2).

In addition, the statute gave blanket authority to the South Carolina Public Service Commission (SCPSC) to help ensure unrestricted cost recovery for participating IOUs: "(D) The commission [SCPSC] is authorized to promulgate, in accordance with the provisions of this section, *all regulations necessary to allow the recovery by electrical utilities of all their prudently incurred distributed energy resource program implementation costs* incurred pursuant to Sections 58-39-130 and 58-39-140 of this chapter [emphasis added]" (South Carolina Senate Bill 1189/Act 236 2014, §2).

The following exchange with a different member of the EAC, representing environmental and renewable energy interests, is illustrative of IOUs' influence in the negotiations over SB 1189:

> **Interviewer:** "How did it work in terms of the nuts and bolts? Because, like, the utilities, like, they would still get full cost recovery?"
>
> **Interviewee:** "Well, *that's why they were okay with [it]* . . . ultimately, like, kind of where this landed is they got lost revenues for pretty much everything they did [emphasis added]." (author's interview, 5/19/22)

A staffer on one of the legislative committees that worked on SB 1189 recalled:

> The Energy Advisory Committee [sic], which I helped staff, I think it was formed in 2011, I want to say. And it was a big committee. . . . And we met in Columbia a lot . . . But then we went to Wild Dunes, you know, I mean, which is not some five-star resort but, you know, it's an enclosed—you know, just locked them in a room, everybody, for two days and met on policy. And, you know, *the utilities wrote that bill* [SB 1189], there's some concessions in there that got rooftop

> going [and] that created some procurement of utility-scale [solar]. *But that bill was drafted by Duke and at the time SCANA*, and that was because they had influence back then [emphasis added]. (author's interview, 5/18/22)

The same staffer went on to note: "Remember, we're starting from this whole philosophy of making the utility whole, which was the bullshit back then" (author's interview, 5/18/22).

In South Carolina, the IOUs accepted SB 1189/Act 236 *only* because they negotiated its content so that they would be protected from any costs/risks associated with their voluntary participation in an RPS-like program. And it likely became law only because they accepted it. Regardless of the environmental potential of the voluntary program, from a distributive politics perspective, IOUs were the clear winners. In the next chapter, we will explore why IOUs seem to be unusual—even among business actors—in their ability to win so consistently.

Oklahoma

In the Oklahoma case, the voluntary RPS legislation, HB 3028, was completely symbolic and nonprescriptive, so there was not even a need for IOUs to worry about cost recovery because there were no specific programs spelled out in the legislation, as was the case in South Carolina. In addition, readers will recall from chapter 4 that a couple of features of the Oklahoma case made it unusual. First, Oklahoma—in contrast with South Carolina—was part of an RTO, the Southwest Power Pool. Second, there was a restructuring proposal (which ultimately failed to be enacted) *initiated by the IOUs* to establish a "deregulated monopoly," in the words of several policymakers who were involved at the time (author's interviews, multiple).

The Oklahoma case is also unusual among the seven because rather than the IOUs playing political "defense," by crafting policy that *others initiated* to ameliorate potential costs/risks to themselves, in this case, the state's vertically integrated IOUs, OG&E and PSO, actually went on the "political offense" by attempting to take advantage of the restructuring wave going on throughout the country to gain an even better status quo for themselves. On the one hand, Oklahoma IOUs' lack of success on "offense" illustrates that there are certain limits to IOUs' political influence.[6] On the other hand, as the Oklahoma IOU lobbyist from the previous chapter noted, this unsuccessful effort left them no worse off than the very strong position they were already in. We will briefly examine the motives behind this effort because it is illustrative of the approach IOUs tend to take to electricity policymaking more generally.

The IOU-initiated restructuring proposal, SB 220, explicitly *did not* require generation divestiture, which readers will recall from the Pennsylvania case in the previous chapter is a pro-IOU and anti-consumer, anti-IPP, policy design principle. Section 6 read: "Nothing in this section shall require an entity to divest itself of any generation, transmission, or distribution assets" (Oklahoma Senate Bill 220 2000, §6). Section 8 of SB 220 was also pro-IOU and anti-consumer, as well as anti-IPP, because it allowed for what industry insiders refer to as "self-dealing." That is, it set up a "competitive" bidding process for wholesale power in which an IOU in the distribution business could contract with its own affiliate (as opposed to a competitive IPP) for generation. That section read, in part: "Every electric distributor, or its affiliate, shall use a competitive bidding process established pursuant to rules promulgated by the oversight and supervision authority. . . . *An affiliate of an electric distributor shall not be prohibited from participation in the competitive bidding process* [emphasis added]" (Oklahoma Senate Bill 220 2000, §8(A)(2)).

A former utility regulator in Oklahoma explained why this IOU-backed restructuring proposal would not have worked for other energy system actors besides the incumbent IOUs themselves: "Independent power producers, you know, that weren't utility-owned, would try to build independent power, you know, generation, in Oklahoma and the investor-owned utilities would fight like crazy to keep them off the wires. So, how the heck are you gonna ever have, you know, competition? Because I was seeing it as someone who was looking at the rate cases. I'm going, 'Okay, these guys are gonna just be deregulated monopolies'" (author's interview, 3/6/23).

The degree to which this arrangement would have been bad for electricity *customers* (who, as chapter 4 points out, were already paying very low rates relative to those in other states) was articulated by a lobbyist representing them, in this case representing *industrial*—big business—customers: "We were supportive of it [restructuring] in principle. . . . Then when you got to the actual drafting of the bill, it took a turn for the worse. . . . And it ended up being, rather than a pro-competition bill, in our opinion, an anticompetition bill and basically a utility protection bill" (author's interview, 2/22/23).

Although restructuring in Oklahoma was never ultimately enacted, a former state legislator's recollection of the politics, including *how close* it got to becoming the law, is illustrative of the influence that IOUs enjoyed at the Capitol in Oklahoma City:

> One of the big issues out there while I was there [serving in the legislature] was deregulation [also known as restructuring] of the utilities. . . . What I remember specifically was there were a lot of lobbyists in the Capitol. . . . What I remember most about the deal was the bill came through, I think it excelled through the House the first time because they removed the title. . . . So, in Oklahoma, our constitution says the bill must have a title and an enacting clause. And striking the title means we can pass it in both houses. But if it was to go to the Governor's desk without the title on it, even if he signed it, it wouldn't become law. So one of the bodies has to restore the title. . . . So I remember that bill coming up. I remember they were working that bill quite a bit. And I think it had quite a bit of support. Then they started having all the rolling blackouts in California. It had quite a bit of support when that session started, but as we moved through and California started having all those blackouts, we started hearing from more and more constituents about not wanting deregulation. (author's interview, 3/1/23)

In the Oklahoma case, cost/risk-shifting within the context of the RPS was a moot point. And yet, even outside of this context, vertically integrated IOUs aggressively sought, and almost succeeded in creating, a policy that would have further cemented their advantage over IPPs under the guise of "competition," which had been concurrently quite popular, and indeed implemented as part of the electricity market liberalization era in many other states in the country.

Oregon: Vertically Integrated IOUs Conditionally OK the RPS

Readers will recall from chapter 3 that political economy factors in Oregon in 2007 were extremely favorable when that state first adopted its mandatory RPS. With an unmatched ability to *shape* legislation but not necessarily a guarantee of always being able to completely *stop* it, Oregon's major IOUs, Pacific Power and Portland General Electric, "saw the train coming down the tracks," according to one Salem lobbyist (author's interview, 2/14/18), meaning that they saw RPS passage in *some* form as being "pretty inevitable" (author's interview, 3/23/18). Readers will recall from the opening of chapter 1 that the political context at that time included a recently adopted RPS by *citizen ballot initiative* in Washington State, Oregon's neighbor to the north, whose design Oregon IOUs very much disliked. As an Oregon IOU lobbyist explained: "Basically, we system plan. All the investments that we make are for the benefit of the whole system, not one state. To boil it down, the [Washington] ballot measure [if passed in Oregon] would've put us in a position where we were building a portfolio exclusively for Oregon, which would've been incredibly expensive" (author's interview, 2/26/18).

With Oregon environmental advocates "threatening that [ballot initiative] here [in Oregon]" (author's interview, 2/14/18), Oregon IOUs, led by Pacific Power and reluctantly joined by PGE, which at the time still owned one in-state coal-fired power plant, decided that their best option was to offer support conditioned on having a prominent—perhaps *the most* prominent—seat at the table in negotiating the bill's language. A high-level staffer for Oregon's governor at the time, who was in the room as much of the negotiating was taking place, described the IOUs' calculus in this way: "In knowing that this green stuff is coming, can we get ahead of it and get some certainty, and some things that we wanted in the legislation?" (author's interview, 2/16/18).

From the governor's perspective, having the political support of the IOUs, however conditional that support was, was indispensable for the goal of overcoming opposition from what I have called elsewhere "traditional business" actors (Basseches 2024), meaning those generally opposed to climate and renewable energy policy action due to concerns about costs. A top climate and energy staffer for the governor, who Rosenthal (2004, 165) refers to as a state's "chief legislator" due to their formal and informal agenda-setting powers, explained the importance of having the IOUs on the same side as environmental advocates and "green business" actors (as opposed to being on the side of the industry forces opposing the RPS) to get the bill passed: "The key was the IOUs came on board. . . . It was one or two votes at the end of the day, *even with* the utilities on board. . . . It was touch and go all the way through the negotiations over how much the utilities were going to water it down, what they were going to get as their benefits [emphasis in original]" (author's interview, 2/16/18).

An IOU lobbyist recalled, from his perspective: "I remember having a conversation with 'Mark'[7] [a staffer in the governor's office], and I can't remember the exact timeline but it was probably early fall 2006, where he was very clear that this [the RPS] was a top priority for [Governor] Kulongoski, that they would be bringing legislation, and that they really wanted us to help them, to partner with them to get something crafted that could pass the legislature" (author's interview, 2/26/18).

In Oregon, even vertically integrated IOUs were ultimately willing to lend their support to a mandatory RPS in 2007, making Oregon the only non–fully restructured state examined in this book that adopted a mandatory RPS, albeit later in time than any of the other mandatory RPS-adopting states, which *were* fully restructured.[8] It was helpful that despite being vertically integrated, with the exception of PGE's Boardman coal plant, the IOUs did not own in-state coal generation assets and instead relied heavily on hydroelectricity.

However, Oregon IOUs' support came with a litany of conditions and strings attached, and Oregon's IOUs were inarguably *the* extralegislative actor with the single greatest say over what the language of that policy looked like. As one Oregon IOU lobbyist put it, his objective was "to figure out how to put in provisions that ameliorate risk while at the same time accomplishing the overall policy" (author's interview, 2/19/18). This goal was successfully accomplished, in the initial 2007 RPS law as well as in subsequent updates to the policy.

There are numerous provisions that could be used as illustrations of IOUs' unparalleled influence in shaping the distributive effects of the Oregon mandatory RPS policy, some others of which are described elsewhere (see Basseches 2024), but I will focus here on four in particular: 1) securing an "automatic adjustment clause," 2) securing the unlimited ability to bank renewable energy credits (RECs) used for compliance, 3) clarifying REC ownership to reside exclusively with the IOUs, and 4) securing accelerated depreciation as a means of complying with the 2016 RPS update. I will discuss each one, in turn.

Section 13 of Senate Bill 838, the initial 2007 Oregon RPS law, states: "The Public Utility Commission shall establish an automatic adjustment clause . . . that allows timely recovery of costs prudently incurred by an electric company to construct or otherwise acquire facilities that generate electricity from renewable energy sources and for associated electricity transmission" (Oregon Senate Bill 838/Act 301 2007, §13). Under preexisting law, making upward adjustments to customers' electricity rates required the IOUs to undergo a "rate case," during which the IOU was required to disclose its finances for review by the PUC and any other interested parties, such as public interest groups. Rate cases are typically lengthy processes that expose the IOU to appropriate levels of scrutiny. In contrast, the "automatic adjustment clause" secured in Senate Bill 838 allows for bypassing this normal process for RPS-related costs, allowing such costs to be passed through to ratepayers quickly and with minimal scrutiny.

A ratepayer advocate in Oregon described the typical rate case, in contrast to the single-issue ratemaking newly available to IOUs under the automatic adjustment clause:

> A rate case is usually about a ten-month process that starts with the utility filing an application. It's in total—I don't know, it's about twelve inches, fourteen inches thick. And it comes with a full set of testimony, written testimony by their witnesses identifying the new revenue requirement they are looking for. . . . So, in establishing that revenue requirement, it's important to determine how they're doing financially, whether they're able to actually earn a return on their investment and all that kind of stuff. . . . And then there are stages of testimony. So the next stage would be [PUC] staff and all intervenors [i.e. other parties seeking to contest the amount the IOU is seeking]. (author's interview, 2/9/18)

IOUs in Oregon were looking for ways to quickly add to the infrastructure costs that would be included in the rate base, thus potentially increasing the return on equity that could be claimed for shareholders as a percentage of it. Of course, this was not the argument they used with policymakers. A PUC official summarized their argument to policymakers in the following way:

> You're talking about an era where the cost of some of these resources at the time could be argued to be more expensive than a combined cycle natural gas combustion facility. And so the utilities I think probably were—felt a little bit caught in a little bit of a bind. . . . This was the first time that the legislature was saying. . . . We are telling you that you need to include this amount of this particular kind of resource. So I think the utilities were a little bit nervous. (author's interview, 3/28/18)

Surely, if captive ratepayers had understood at the time that this legislation was offering monopoly utility companies a means of raising their rates *without* the usual level of scrutiny, they too would have been at least "a little bit nervous." But due to the unique degree of political influence enjoyed by the vertically integrated IOUs (Basseches 2024), *their* nervousness was assuaged through provisions like this one. As an official in the Kulongoski administration put it, characterizing the nature of the IOUs' participation in the 2007 negotiations around the RPS statute's language: "It was in their interest to see what they could get. . . . And they had some issues around regulatory treatment. They didn't want to have to go through a whole new rate case to get wind into their portfolio. . . . We created a path [i.e., the automatic adjustment clause] where you could go and do a modified thing" (author's interview, 2/26/18).

A second official put it more bluntly: "That [automatic adjustment clause] was a concession to the utilities to get them on board" (author's interview, 2/27/18). When the RPS was updated legislatively in 2009 to include a solar pilot program, section 3 of the bill made clear that the automatic adjustment clause would apply to that program as well (Oregon House Bill 3039/ Act 748 2009, §3(5)).

Another important provision to Oregon IOUs for ameliorating potential costs or risks was the ability to bank RECs indefinitely, for use in future compliance periods. This means that rather than adding the specified amount of capacity per compliance period, an IOU could instead invest heavily in capacity in early compliance periods and then potentially not at all in later ones, using excess/banked allowances accumulated in the early periods as a means of demonstrating compliance to regulators in the later ones. SB 838 reads, in section 16: "Renewable energy certificates that are not used by an electric utility or electricity service supplier to comply with a renewable portfolio standard in a calendar year may be banked and carried forward indefinitely for the purpose of complying with a renewable portfolio standard in a subsequent year" (Oregon Senate Bill 838/Act 301 2007, §16(2)).

When it came to distributive politics, this provision pitted the interests of IOUs, who wanted maximal flexibility in terms of when and how they added renewable energy capacity, against those of independent power producers, who preferred the assurance that contracts for their clean electricity would be pursued with regularity and dependability. A PUC official described the banking rules that ultimately prevailed as "very liberal" (author's interview, 3/28/18), meaning favorable to IOUs. A staffer to the governor described them as "good for the utilities" (author's interview, 2/26/18), and a state legislator said the goal was to offer IOUs "more flexibility with the credits" (author's interview, 2/8/18). By contrast, a leading "green business" lobbyist (Basseches 2024), who reluctantly acquiesced to the IOUs during the negotiations for the sake of securing legislative passage, described the provision as a "bad idea" and something she wished "we could undo" (author's interview, 2/16/18).

When the Oregon RPS was updated legislatively in 2009, pilot programs were established to promote behind-the-meter solar PV adoption by customers in the IOUs' service territories. To avoid any ambiguity over who would own the title to the associated RECs—the customer versus the IOU—language was added stating: "Ownership of renewable energy certificates established under ORS 469A.130 that are associated with renewable energy generation that is sold to an electric company under the pilot programs must be transferred to the electric company and may be used to comply with the renewable portfolio standard described in ORS 469A.052 or 469A.055" (Oregon House Bill 3039/Act 748 2009, §2(8)).

The distributive implications of this provision are clear. Although, under the new pilot programs, consumers were encouraged to self-generate renewable electricity, the monetary value of the associated RECs belonged not to them but to the IOU in whose service territory they happened to reside. Furthermore, even though it is the *utility*, rather than the customer, who has compliance obligations under the RPS, the utility—in addition to being able to pass any RPS-related costs through to the customer base vis-à-vis mechanisms such as the automatic adjustment clause—was allowed to also fulfill its obligations using electricity generated not by its own generation facilities nor by those of IPPs it contracts with, but by its captive customers.

Finally, when the RPS was updated legislatively in 2016 to mandate for the first time ever the complete phase-out of in-state, coal-fired power from Oregon IOUs' portfolios, it accommodated the one IOU, PGE, that still owned an in-state coal-fired power plant, by directing the PUC to automatically approve a schedule of accelerated depreciation of that asset to ensure that the company's shareholders would in no way be adversely affected. Section 1 of that law states: "The Public Utility Commission shall adjust any schedule of depreciation approved by the commission for an electric company's coal-fired resource . . . [such that] the adjusted depreciation schedule . . . require[s] the coal-fired resource to be fully depreciated on or before December 31, 2030" (Oregon Senate Bill 1547/Act 28 2016, §1). Once again, we see an example of a provision that reduces costs to IOU shareholders, despite the arguable imprudence of the company's past investment decisions, at the expense of other stakeholders.

Texas, Pennsylvania, California, and Massachusetts: IOUs Shape *Both* Restructuring *and* Mandatory RPS

As argued in chapters 3 and 4 of this book and elsewhere (see Basseches 2024), state policies that fully restructured the electric utility sector created feedback effects that altered IOUs' policy preferences in a way that made some of them much more favorable to subsequent RPS policies, and others of them who would oppose or seek to water the RPS policies down less politically influential than they would have been had they remained vertically integrated monopolies. Such was the case in Texas, Pennsylvania, California, and Massachusetts, all of which were among those states that adopted restructuring policies in the 1990s as part of the electricity market liberalization era.

In this section, I will show that even the IOUs in these fully restructured states were consistently winners when it came to distributive politics. First, they helped shape those initial restructuring policies in all four cases. Second, those postrestructuring IOUs that were responsible for demonstrating compliance under the state RPSs (i.e., transmission and distribution utilities, and retail electric providers in the Texas case) still managed to shape the RPS policies in ways that shifted costs/risks from themselves onto others. I focus primarily on Texas and Pennsylvania here, since California and Massachusetts are discussed elsewhere (see Basseches 2024).

In both Texas and Pennsylvania, the most important way in which IOUs shaped the distributive effects of the restructuring policies was their treatment of "stranded costs," which are defined in chapter 3. In both states, a per-utility stranded cost value was ultimately approved by the state PUC, a venue in which we know IOUs enjoy great influence (Heern 2023; Hirsh 1999). In both states, this stranded cost value was to be passed through to ratepayers via a "non-bypassable competitive transition charge" (CTC). This meant that the CTC could not be avoided by ratepayers, regardless of whether they chose to take advantage of a competitive supplier once retail competition was in effect. However, the Texas legislation *did* offer an additional consumer protection measure, in an attempt to safeguard against the "over-recovery" of such stranded costs; the Pennsylvania legislation, which preceded it by three years, did not.

SB 7 (adopted in 1999) added chapter 39 to the Texas Public Utility Code, which, in section 252, established a utility's "right to recover" "all its net, verifiable, non-mitigatable stranded costs incurred in purchasing power and providing electric generation service" (Texas Public Utility Code, chapter 39, §252). The Texas statute was arguably more consumer-friendly than the Pennsylvania one, because it also included section 262, a "true-up proceeding," in which the Texas PUC would require the utility to reconcile any differences between its *estimated* stranded cost value, which was the basis of the CTC amount, and its *actual* stranded costs, calculated at a later date according to methodologies prescribed in the statute. Any overpayment on the part of ratepayers, based on the IOU's initial estimate, was to be credited back to the ratepayers (Texas Public Utility Code, chapter 39, §262).

Although provisions like this one, according to one former Texas state senator, contributed to IOUs' initial dislike of the legislation (author's interview, 3/12/22), the general sense among those interviewed was that the IOUs were pleased with the stranded cost provisions outlined in the legislation, and ultimately acceded to the legislation as a result. As one lobbyist involved in the negotiations put it: "At first, the utilities were against it [the restructuring bill], but then they thought, 'Maybe we can make more money' and they thought they could control the process. . . . [On] the stranded cost issue, they were getting provisions that were really damn good for them, frankly. They even [got] guaranteed recovery. . . . It was a hell of a good deal for them" (author's interview, 3/10/22). A former Texas PUC official recalled: "In the case of the utilities, in '99 [SB 7], they got a whopping amount of stranded cost [recovery], plus securitization[9]" (author's interview, 3/17/22).

Legal scholar Lois Lupica (2000) has argued that the "entitlement" that IOUs convinced policymakers that they had to have 100 percent stranded cost recovery as a condition of restructuring was incorrectly assumed:

> Transition loss recovery advocates have put forth a series of arguments and assertions that, on their face, appear compelling. Many of these arguments, however, are based upon faulty premises and flawed predicate assumptions. Moreover, the electric utility transition loss issue has not been debated in light of the broader societal context in which it belongs. . . . An *ex ante* examination of the electric utility market during the period of [traditional] regulation reveals the utilities' inefficient decision making, which was premised in part upon the flawed assumption of an infinite entitlement to the benefits that flow from regulation. (Lupica 2000, 668, 670)

The stranded cost recovery language, even with Texas's consumer-friendly innovation of the "true-up proceeding," represents one way in which vertically integrated IOUs shaped state policy to shift risks/costs away from themselves and unto ratepayers.

When it came time to make the first and only legislative update to the Texas RPS with Senate Bill 20 in 2005, three notable cost/risk-shifting provisions were included. Two of them shifted costs/risks away from transmission and distribution IOUs and onto ratepayers. A third shifted costs/risks away from REPs and onto "green business" actors (Basseches 2024), specifically competitive renewable energy generation developers.

Section 1 of SB 20 states that, if a transmission and distribution utility is ordered to construct additional transmission infrastructure to facilitate meeting the RPS: "The commission *shall find* that the [transmission] facilities are used and useful to the utility in providing service for purposes of this section and *are prudent and includable in the rate base, regardless of the extent of the utility's actual use of the facilities* [emphasis added]" (Texas Senate Bill 20 2005, §1). The "used and useful" doctrine is a constraint on cost recovery that utility regulators have at their disposal under ordinary ratemaking procedures to deny excessive cost recovery by utilities, for property that is not needed (i.e., not "used and useful") to provide service. Therefore, the legislative language in Section 1 disempowers regulators from reining in the cost recovery of transmission and distribution utilities, so long as those utilities can show that the costs were the result of the RPS.

Relatedly, in section 2 of the same bill, there is a provision that—not unlike the automatic adjustment clause that we saw in Oregon—speeds up the cost recovery process for the IOUs at the expense of commissioners exercising due diligence in reviewing the case, and also at the expense of non-IOU intervenors who might potentially be wanting to raise challenges to the case. Specifically, the language in this section reads that, if a transmission and distribution utility files for recovery for costs associated with RPS compliance: "The commission shall issue a final order before the 181st day after the application is filed with the commission. [And] if the commission does not issue a final order before that date, the application is approved" (Texas Senate Bill 20 2005, §2). This means that if the Commission does not rule to the contrary within 180 days (perhaps due to its limited administrative capacity to apply the necessary scrutiny), the IOU automatically gets its way without any recourse for intervening parties.

In addition, SB 20 contained what one renewable energy advocate described as "a poison pill that was added in" (author's interview, 3/11/22). Specifically, this advocate was referring to the addition of §39.904(m) to the Public Utility Code, which stated: "Notwithstanding any other provision of law, the commission shall ensure that all renewable capacity installed in this state and all renewable energy credits awarded, produced, procured, or sold from renewable capacity in this state are counted [toward the RPS]" (Texas Senate Bill 20 2005, §3).

Although this language might seem innocuous out of context, the context makes it important. The context was the voluntary REC market that was already operating in Texas at the time. By allowing RECs that were purchased *voluntarily* by private parties to count toward the *mandatory* RPS requirement that needed to be met by REPs, this language constituted a de facto weakening of the value of the *mandatory* RECs, thereby reducing costs for the REPs, which are the restructured IOUs that are required to demonstrate compliance under the law. This provision also harmed the business prospects of independent power producers in the renewable energy generation business because it reduced demand for the additional renewable capacity that they were in the business of offering. Although this provision was repealed just two years later,[10] only once it was clear that the 2005 RPS target would be easily surpassed, it nevertheless illustrates a different way in which IOUs were subtly able to shift costs, even in a fully restructured utility sector environment.

In Pennsylvania, the restructuring legislation itself (HB 1509) was less prescriptive than in Texas; more discretion was left to the PUC to determine stranded cost values. It should be noted that a prior version of the restructuring legislation, HB 2300, which died in the House Consumer Affairs Committee and never became law (ultimately being scrapped and replaced by HB 1509), would have been less friendly to the IOUs on the question of stranded costs. The text of HB 2300 read: "A majority of the [stranded] costs shall be primarily absorbed by the electric utility's shareholders and to a lesser extent by the competition transition charge which shall only be applied during the transition period" (Pennsylvania House Bill 2300 1996, §4(13)). However, this was not the bill that became law.

Instead, HB 1509, which *did* become the law, put only the following parameters around the PUC's utility-by-utility stranded cost calculation: "The commission has the power and duty to approve a competitive transition charge for the recovery of transition or stranded costs it determines to be just and reasonable[11] to recover from ratepayers" (Pennsylvania House Bill 1509 1996, §2804(13)). Then, in Section 2808, the statute offers several "principles" that the PUC must apply to its determination of the stranded cost value for each utility. These principles amount to giving IOUs the full value requested, assuming various conditions including that they have made every "reasonable" effort to mitigate (reduce) the amount (Pennsylvania House Bill 1509 1996, §2808(c)).

While analyzing the details of each Pennsylvania IOU's restructuring rate case is beyond the scope of this book,[12] interviews with insiders who were directly involved with the proceedings suggest that the IOUs made out very well in terms of getting their stranded cost estimates incorporated into rates through the nonbypassable CTC. For example, a former Pennsylvania PUC commissioner recalled: "Mechanically how it worked [was that] the law [HB 1509] passed, and then they [each utility] had to file restructuring plans where it was ultimately decided how much stranded cost recovery they got. And you know, of course, the utilities came in with high numbers and the consumer advocates came in with low numbers. And eventually, the PUC, you know, gave the utilities probably 65% to 70% of what they asked for" (author's interview, 4/26/23).

Meanwhile, a Pennsylvania ratepayer advocate recalled his losing battle in the first utility restructuring case at the PUC:

> In the very first case, what we argued was that even if you get stranded costs [recovered], you should not get—there's a difference between return *of* cost and return *on* cost. Return *of* cost is like the depreciation. The return *on* [cost] is like the profit margin. We argued that, okay, you can get to recover your cost, but you shouldn't get a return on it at profit. We lost that argument. We lost that argument at the PUC because the statute didn't clarify that, it didn't say. It just said, "stranded costs" [emphasis in original]. (author's interview, 4/7/23)

In the end, complete stranded cost recovery including the use of securitization as a financing technique was what won over most IOUs in Pennsylvania, leading them to embrace restructuring. As a former Pennsylvania PUC commissioner stated: "My recollection [was that] what really got them [the utilities] on board was when they got an opportunity to earn on stranded costs. And then the big deal that got them on board was that they were able to securitize the stranded cost payments, right? So that meant a big up-front slug of money that they could use based on their stranded cost recovery" (author's interview, 4/26/23).

If the stranded cost and securitization provisions of HB 1509 shifted *costs* away from Pennsylvania IOUs, the "off-ramps" in section 2804 of the same bill shifted financial *risks*, functioning much like a free insurance policy. Section 2804(4)(iii) contains a generous list of "exception[s] to the limitations set forth in subparagraphs (i) and (ii) [i.e., the rate caps during the transition period]." The idea was that, to guard against price gouging during the transition period, regulated rates would be capped until full competition was effectuated. However, this section of the law allowed the utilities to petition for exceptions under a variety of circumstances, such as the utility becoming "subject to significant increases in the rates of Federal or State taxes or other significant changes in law or regulations that would not allow the utility to earn a fair rate of return [with the rate caps in place]," significant increases in fuel costs, being ordered to make infrastructure upgrades, and others (Pennsylvania House Bill 1509 1996, §2804(4)(iii)). The rate caps, which were the ratepayers' end of the restructuring grand bargain, were ultimately allowed to be eroded while the stranded cost recovery, which was the IOUs' end of the restructuring grand bargain, was not. Although, as far as could be ascertained, these "off-ramps"—or this "insurance policy"—was never actually triggered or used, a consumer advocate noted they were important provisions that the IOUs were successful in negotiating for as HB 1509 was being developed (author's interview, 4/12/23).

Similarly, when it came to Pennsylvania's mandatory RPS, the IOUs were highly influential in crafting the policy, shifting the distributive effects in their own favor. One prominent example of this is the force majeure provision of SB 1030. This was a mechanism through which a Pennsylvania IOU could call for the suspension of the entire RPS if compliance were to prove too onerous. Section 2 of SB 1030, defining force majeure, states:

> Upon its own initiative or upon a request of an electric distribution company or an electric generation supplier, the Pennsylvania Public Utility Commission, within 60 days, shall determine if alternative energy resources are reasonably available in the marketplace in sufficient quantities for the electric distribution companies and electric generation suppliers to meet their obligations for that reporting period under this Act. If the commission determines that alternative energy resources are not reasonably available . . . then the commission shall modify the underlying obligation of the electric distribution company or electric generation supplier or recommend to the General Assembly that the underlying obligation be eliminated. (Pennsylvania Senate Bill 1030/Act 213, §2)

Section 3 then states: "An electric distribution company or an electric generation supplier shall be excused from its obligations under this section to the extent that the commission determines that force majeure exists" (Pennsylvania Senate Bill 1030/Act 213, §3). Subsection (e) of section 3, the subsection governing accounting through the PUC-designed "alternative energy credits[13] program," also references the force majeure exception.

Subsequently, in 2008, as the Pennsylvania PUC was crafting implementing regulations pursuant to Act 35, there were further questions raised about exactly how force majeure would work in practice. Environmental advocacy groups as well as "green business" actors (Basseches 2024) argued that the PUC should not be able to declare force majeure "on its own initiative" and that "very narrow and specific circumstances be met before force majeure is declared," outlining twenty-five items in particular (Pennsylvania Public Utility Commission 2008, 23, 26). Clearly, these stakeholders were attempting to minimize the ability for IOUs to access the force majeure relief too readily, thus undermining the efficacy of the RPS. Ultimately, the PUC declined to pursue their recommendations (Pennsylvania Public Utility Commission 2008).

The force majeure provision was a significant win for IOUs. One Pennsylvania IOU lobbyist described it as "a safety valve provision" (author's interview, 5/4/23). Even though his company "never had to go in for it" (author's interview, 5/4/23), it was there just in case. Another IOU lobbyist for a different company described it as a "stop gap" (author's interview, 5/22/23). It reduced risk.

Finally, just as we saw in both Oregon and Texas, Pennsylvania, too, had an automatic adjustment clause provision. This was spelled out in section 3 of SB 1030. A Pennsylvania IOU lobbyist explained: "We got some rate reforms put in place that made it easier to invest, and we got sort of current recovery of your costs instead of having to wait till the next rate case" (author's interview, 5/15/23).

Who Cares Who Pays?

As this chapter has shown, irrespective of whether a state restructured its electric utility sector or not, IOUs were the consistent winners when it came to distributive effects of policy, both the restructuring policies themselves and the RPS policies that followed. As chapters 3 and 4 showed, state-level RPS policies exhibit significant variation when it comes to their environmental stringency—something that has been well-documented in the broader literature (e.g., Carley et al. 2018; Glenna and Thomas 2010; Fischlein and Smith 2013). Yet, when it comes to questions of distributive effects, there seems to be no variation to analyze: in a deeply structural sense, IOUs inevitably emerge as winners nearly every time.

There is a normative question about the degree to which these cost/risk-shifting provisions are problematic. After all, if IOUs—which Stokes (2020) paints together with the fossil fuel industry as obstacles to climate action—can be "won over" to support, or at least not oppose, renewable energy policies, as earlier chapters show they can be, isn't this a good thing in the face of an increasingly urgent climate crisis? If a restructured regulatory environment can incentivize such support, is there even a need to raise questions about who pays for the policies (Basseches 2024)?

I argue that there is, for two reasons. First, there is the importance of a *just* transition, not only a rapid one (Carley and Konisky 2025). Welton and Eisen (2019) argue that, in the absence of proactive just transition policies, the clean energy transition has the potential to exacerbate, not mitigate, economic inequality. In particular, these legal scholars note the importance of the questions of "who benefits [from]" and, on the flip side, "who pays [for]" the clean energy transition. The pattern of cost/risk-shifting provisions secured by IOUs, documented in this chapter, ensures that they always benefit from, and never pay for, the clean energy transition.

Second, if IOUs are always the policy winners at the expense of myriad other stakeholders and interest groups, then this raises serious questions about the quality of democratic representation in this policy space (Chen 2019; Drutman et al. 2019). If our political system is meant to adjudicate among the preferences of competing interest groups, yet one type of interest group consistently wins at the expense of the others, then the system is not functioning as intended. This is related to concerns about regulatory capture (Carpenter and Moss 2014; Rilinger 2023).

On the first issue, within energy justice scholarship, two important tenets are "distributional justice," which Montañés and colleagues (2023) define as "the distribution of (dis)benefits of energy systems," and "procedural justice," which they define as "the inclusivity and representativeness of decision-making practices in energy policy" (Montañés et al. 2023, 438). IOUs' consistent ability to shift costs/risks onto ratepayers and/or "green business" actors most obviously implicates distributional justice, but it also implicates procedural justice, because of the asymmetry in participation, in some cases, and influence, in all cases, between IOUs and competing stakeholders in policymaking deliberations, whether in state legislatures or at the PUC. Indeed, Carley and Konisky (2020) note that suboptimal "citizen involvement in U.S. [state-level] utility commissions and Federal Energy

Regulatory Commission decisions" is an important example of a procedural justice concern related to the clean energy transition (Carley and Konisky 2020, 572). My research finds that, while there is variation across states in the quality of non-IOU stakeholder *participation* in such proceedings, their outcomes—at least regarding the implementation of RPS policies—most often favor utility interests.

On the second, much of the interest group literature in political science is premised on the ideal of "interest group pluralism"; that is, while it may be unrealistic to expect ordinary citizens to actively participate in policymaking processes, the interest groups that represent them should at least have relatively equal voice (Truman 1955). Yet, more recent research has suggested this is not the case; there is a "top tier" of interest groups that enjoy superior representation (Drutman et al. 2019), with one key mechanism being how information that decision-makers receive is filtered through their staff (Hertel-Fernandez et al. 2019; Henderson et al. 2023). There is an underlying assumption in this work that interest groups are competing with one another for policy influence (Hacker and Pierson 2014). And yet, in this particular space of electricity policy, the highly unusual "regulated monopoly" structure of electric utilities creates a strange dynamic where, despite being *corporate* nonstate actors, IOUs seem to function as almost a privatized extension of the state, influencing its policymaking in a singular way in which the IOU monopoly seems to *always* benefit, even as others attempt to intervene (Basseches 2024). Certainly, restructured states mitigate this by limiting monopoly control to the transmission and distribution functions of the electricity system, yet the effect still seems to persist at that level. In the next chapter, we will explore further the mechanisms of this most unusual form of political power.

6 Lessons and Implications

This book has brought new attention to the importance of IOUs, and utility sector structure and regulation more generally, in the study of state-level climate and renewable energy policymaking. Motivated by a desire to explain variation in state-level RPS policy design at the more granular "subpolicy level" (Parinandi 2023), my research analyzes the critical importance of partisanship (and associated ideologies) and state energy economy. However, at the same time, it points to this third major and understudied factor—utility sector structure—and shows how it is critical to the interest group politics that inform these policies. Beyond that, and following Stokes (2020), my research highlights the importance of policy feedback effects (Pierson 1993). Specifically, I show how the era of "electricity market liberalization" (Lee 2020) and state-level restructuring policies reshaped the interest group politics of state-level RPS policies, by dividing the policy preferences of IOUs in fully restructured states. At the same time, as the previous chapter makes clear, on the dimension of "who pays?" (Basseches 2024), IOUs emerge as the clear and consistent winners regardless of utility sector structure.

In chapter 1, I reviewed the history and governance structures of the US electricity system to show how understanding these things is essential for making sense of divergences in RPS policy regimes. In chapter 2, I introduced the research questions and theoretical perspectives that guided this inquiry, as well as the data and methodological approach utilized to analyze the politics of RPS policymaking in seven US states. In chapter 3, I showed how partisanship and associated political ideologies explained differences between Oregon's mandatory and South Carolina's voluntary RPS regimes, but how utility sector restructuring was necessary to understand the political feasibility of the mandatory RPS outcome in Texas. In chapter 4, I showed how the fossil-fueled economy of Oklahoma best explained its purely voluntary RPS but also how a fully restructured electricity sector in Pennsylvania enabled its mandatory RPS, despite the fossil fuel industry largely accounting for the limitations of its design. Finally, in chapter 5, I turned to the distributive politics of electricity policymaking and showed how they favor IOUs irrespective of utility sector structure.

In this concluding chapter, I begin with a deeper exploration into why IOUs are such politically dominant actors in the electricity policymaking space. This enables us to understand why *their* policy preferences regarding RPS policy design were so determinative of policy outcomes across the board. I then advance some recommendations for democratizing policymaking in this space.

The chapter then turns to a discussion of the lessons and implications of these findings for the academic literature on state-level climate and renewable energy policymaking. Now that we know how important utility sector structure is, and how politically influential IOUs are, how should that change the way we think about these policies? Finally, what directions for future research do these findings point to?

Sources and Mechanisms of IOUs' Unmatched Political Influence

There is a large, interdisciplinary literature on "regulatory capture" that is useful to better understand the nature of IOUs' distinct form of political influence. That literature can arguably be traced back to Stigler (1971). "Material capture," which in more layman's terms is often called "corruption,"[1] is infrequent, and thus regulatory capture conceived of in this way is exceedingly difficult to detect, measure and prevent (Carpenter and Moss 2014). Carpenter and Moss (2014, 12) distinguish between "strong capture" and "weak capture," arguing that the latter is far more common and requires a lower burden of proof—only that "special interest influence compromises the capacity of regulation to enhance the public interest."

McCarty (2014) argues that there is greater potential, at least for "weak capture," in policy areas that are more technically complex, and certainly electricity policy would be an example of this. The more technically complex the policy area, the greater the value of the "legislative subsidies" (Hall and Deardorff 2006) that outside actors such as IOUs can provide. In addition, as regulators themselves seek to wrap their heads around these technically complex issues, there is a greater chance of "cognitive capture," in which regulators, without malicious intent, simply come to adopt the perspective of those they regulate as their own (Rilinger 2023).

In previous work (see Basseches 2024), I identified four possible reasons for the unmatched influence of IOUs in the electricity policy space: 1) the size of these companies, 2) the skill of their lobbyists, 3) the technical expertise they possess, and 4) the flexible corporate forms in which they are organized. In the following sections, I elaborate on the second and third of these reasons, as well as address the other "elephant in the room" when it comes to their influence: They are private monopolies providing an essential public service.

Lobbyists/Consultants/Lawyers and the Technical Knowledge They Bring

In all seven states, regardless of utility sector structure, the IOUs tended to have the advantage when it came to lobbyists, lawyers, and paid consultants—whether in-house or contract. In the context of *legislative* lobbying, LaPira and Thomas (2017) discuss the distinct advantages offered by contract versus in-house lobbyists, and, as I have pointed out elsewhere (see Basseches 2024), IOUs more often than competing actors are consistently armed with both, creating a "best of both worlds" dynamic when it comes to realizing their preferences in state legislatures. As one California environmental advocate described it: "They [utilities] have got lobbyists up the wazoo" (author's interview, 10/30/18). From the perspective of a South Carolina state legislator:

> There's a natural tendency when you're dealing with something very complicated or complex to defer to the experts, right? And the experts in this matter [i.e., electricity policy] are, obviously, the utilities, right? The utilities are the ones that do this. And so they're the ones that have the extremely talented executives, they have the very, very talented lobbyists. They speak very knowledgeably about the subject matter. And look, it's very intimidating if you're a legislator and you're trying—I mean, you've got a lot on your plate, you know, dozens of public policy areas that you have to try to master. There's a natural tendency to defer to the expert. (author's interview, 6/1/22)

This particular legislator recognized that the information received from utility lobbyists was not objective, but it was also clear that it was hard to find alternative sources that were. The same legislator continued: "And so part of this process is educating ourselves and getting ourselves to the point where, as legislators, we can make decisions and we can enact policies that are based on what we perceive to be in the best interests of South Carolinians. And that isn't necessarily what is being told to us by the experts [read: utilities], right? Because they have a vested interest in this" (author's interview, 6/1/22).

Meanwhile, at the PUCs—the other venue in which IOUs have a major stake in policy design and outcomes (Heern 2023; Parinandi 2023)[2]—the dynamic is different but still advantageous to IOUs relative to competing actors, such as IPPs, environmental groups, or consumer advocacy groups. In the case of the PUCs, despite variation in prior professional experiences of PUC commissioners (Heern 2023), the commissioners are typically more learned on the issues than legislators, since their policy purview is narrower and more specialized.[3] Yet even in this case, there is an asymmetry of influence, since the IOUs are the actor most likely to be armed with the best-resourced regulatory attorneys and consultants, again both in-house and on a contractual basis. A Pennsylvania ratepayer advocate in PUC proceedings put it this way:

> I mean, the utilities always have an advantage going into any case [at the PUC]. They have all the information, you know, we don't. Pennsylvania has a very short timeframe for its cases. For the restructuring cases, we had the same nine-month timeframe as we do for a base rate case. They [utilities] obviously have unlimited resources in terms of what they can financially spend, as well as the folks that are working there. You know, our office had to hire experts, and we were fortunate that we were able to hire, you know, excellent expert witnesses, but they had to do immense, massive discovery to be able to get the information they needed to do the case. So the utilities always had an advantage. (author's interview, 4/12/23)

In the case of every IOU across every one of the seven states, the companies had separate but coordinated departments handling government affairs (i.e., lobbying) in the legislature and regulatory affairs at the PUC, and they would make strategic choices about whether to lobby the legislature to maximize the policymaking discretion of the PUC (in cases in which they determined they had better odds of exerting influence there) or, conversely, whether to limit the PUC's authority through legislation. Their typical preference seemed to be the former, which also tended to make legislators, who felt less confident in their own expertise, more comfortable voting for bills that delegated decisions to the PUCs (author's interviews, multiple).

Of course, it is not *just* the *human* resources (in the form of lobbyists, lawyers, and consultants) that IOUs bring to the table that contributes to their influence over public policy; it is also the underlying technical complexity of the electricity system itself, which they themselves first designed and stood up, and have maintained for more than a century since (Hirsh 1999). Virtually everyone knows how to turn on a light switch, but far fewer people understand all that must happen to ensure the physical/mechanical, financial, and legal integrity of the overall system that guarantees that, when that light switch is turned on, the electric supply will be there for the room to become illuminated. The IOUs know it the best, since it's their system. Policymakers do not always even know what data to ask for from them.

Sometimes, policymakers have *too much* data rather than not enough, and lack the ability to interpret it all, or to pick out what is relevant for a particular policy question. An IOU might flood a regulator with so much information that it is difficult for the regulator to know where to even begin in analyzing it. This problem is reflected in the comments of a Texas PUC commissioner, who told an energy policy podcaster: "In my words, what I would say is regulators have at their disposal a huge amount of data. Regulators have at their disposal very little information" (Lewin 2025).

From my research, the following quotes, from three former PUC or state-level energy agency commissioners, in Massachusetts, California, and Oregon, respectively, are illustrative of this asymmetry in technical knowledge and expertise between IOUs and their regulators. Keep in mind, Massachusetts and California are restructured states while Oregon is vertically integrated:

> The utilities always have a very dominant role to play because they own and operate the system and know how it works. When they essentially say, "Such and such cannot work, it cannot be done," that tends to make it a pretty impossible thing to get done.
>
> —Former Massachusetts energy commissioner (author's interview, 8/20/18)

> I felt like I was always looking at an individual decision, trying to make the best decision I could. But I knew the utilities internally were tracking much better [than the PUC] what the overall picture looked like.
>
> —Former California PUC commissioner (author's interview, 7/16/19)

> It's really hard for the Commission, or anybody else, to go up against the technical power of the utilities.
>
> —Former Oregon PUC commissioner (author's interview, 3/2/18)

The "Elephant in the Room": Private Monopolies Providing an Essential Public Service

Given the dynamics described earlier, there is often a blurring between utility companies and the state itself. After all, the laws governing such companies are referred to in statute as the "Public Utility Code," yet the utility companies themselves are, of course, not actually public; they are, in fact, private, monopoly corporations. They are public only in that they provide a *public service*, much like the state itself often does with, for instance, roads, postal service, public safety, and so on. In other words, it is not just that IOUs are powerful in an instrumentalist formulation of power (Miliband 1969), meaning that their lobbyists/lawyers/consultants are often the "smartest" in the room and have the greatest ability to direct policy, but also in a deeply structural formulation of power (Poulantzas 1973): The interests of the *state* become difficult to differentiate from the interests of the *utility*. And yet, only the utility, not the state, is motivated by shareholder profit.

There are two particularly vivid examples of this tight coupling between the state's and the private monopoly utility's interests. The first is how the State of California went to great lengths, at its taxpayers' expense, to keep its private monopoly investor-owned utilities as financially healthy as possible, both in the wake of that state's energy crisis in 2000–2001 and in the face of their liability for deadly wildfires induced by their power lines in the late 2010s. The second is, more generally, the function of state attorneys general, which ostensibly act as ratepayer advocates, but at the same time also have an interest in maintaining the financial health of the investor-owned utilities. I will discuss each of these, in turn.

In 2000–2001, as California was implementing restructuring, the state entered a well-documented "energy crisis" in which wholesale power prices spiked due to market manipulation and market design failures (Rilinger 2024), and at the same time, the transmission and distribution IOUs were prohibited from raising retail rates by the state's 1996 restructuring law. The results included widespread blackouts, bankruptcy for Pacific Gas and Electric (PG&E), and ultimately, the defeat of the governor, Gray Davis, by Arnold Schwarzenegger in a 2003 recall election (Blunt 2022).

Both PG&E and Southern California Edison (SCE) lobbied Governor Davis (and the PUC he controlled through his political appointments), as well as the legislature, to allow them to raise retail rates. The utility in the most immediate trouble was PG&E, and Governor Davis stood firm and said no to a rate increase. As a result, PG&E declared bankruptcy. SCE was next, but this time, the governor intervened, perhaps because his hand-picked PUC president was formerly that company's CEO (Blunt 2022).

PG&E would eventually find its way out of bankruptcy, by selling off the holding company's nonregulated businesses[4] and by getting the California PUC to agree to a proposal that would have the company "reduce rates in the near term but collect from customers as much as $8 billion in extra revenue over the next decade" (Blunt 2022, 83). One of the PUC commissioners involved in the deal recalled: "The biggest thing, though, was to get PG&E—make a settlement with PG&E and get them out of bankruptcy, to get them back to being a functioning utility in the State of California. Because the state's energy goals are very difficult to meet when someone's in bankruptcy, the biggest utility in the state" (author's interview, 7/3/19). No good deed went unpunished for Governor Davis, who had the audacity to say no to PG&E when first approached for a bailout. Davis paid the ultimate political price for trying to look out for the interest of California's ratepayers at the expense of its IOUs, losing in a recall election to Arnold Schwarzenegger (Blunt 2022).

Not only was the *state* dependent on its regulated IOUs being in good financial health, but ironically, so too was a type of business actor that frequently has conflicting policy preferences with the IOUs when it comes to renewable energy policy (Basseches 2024): independent power producers (IPPs). A lobbyist representing IPPs explained: "We've been pretty strong advocates on behalf of the independent power industry. . . . So this has brought us into, oftentimes, conflict with the utilities. . . . It's a little awkward in that they are, on the one hand—commercially, there's some adversarial interests there, but we're also commercially dependent upon them because they buy our product" (author's interview, 8/23/19).

About a decade and a half later, PG&E—the same company that went into bankruptcy because of the energy crisis of 2000–2001—entered bankruptcy for a second time. This time, the cause was California's "strict liability" for damages due to wildfires caused by utility equipment. During another tumultuous few years, several options were considered to deal with the utility's financial challenges. These included municipalization, whereby San Francisco would acquire some of the company's assets and run a portion of its transmission and distribution system (Bade 2019), splitting the company into separate gas and electric companies (Walton 2019), and even amending the state law to limit utilities' strict liability (Trabish 2019).

In the end, the issue was resolved in a way not all that different from how PG&E's first bankruptcy was resolved: a bailout. Governor Gavin Newsom signed into law AB 1054, which created a $21 billion taxpayer-funded account to help utilities pay off their wildfire-related debts. The *Los Angeles Times* Editorial Board called the legislation "badly vetted." A public interest advocacy group called the measure a "bailout" (Walton 2019a). Although months earlier PUC regulators had reportedly stated that a bailout was "politically untenable" (Walton 2019b), it seemed to be exactly what ultimately happened. Upon his resignation, the California PUC President "wondered aloud whether PG&E is now too big to regulate" (Walton 2019c).

A second example of the structural power of IOUs has to do with the official role of state attorneys general when it comes to intervening in PUC cases on behalf of the public interest. In most states, the attorney general has the power and/or duty to intervene in utility rate cases to protect the public interest, under the state constitution, statute, or common law (National Association of Attorneys General 2024). Yet, because IOUs provide an essential public service, and because there is often no plausible alternative provider of that service, the divergences between their interests and those of the public are often not easily discernible. This tension is reflected in the data examined for this book, including the interviews. For example, a former staff member in the Oklahoma attorney general's office stated: "The statute actually says that the AG is to represent the interest of ratepayers. But the twist on that, and I've said this to several people, was from my perspective . . . was that part of doing well for consumers is to have healthy utilities. Not fat utilities, but healthy utilities" (author's interview, 3/2/23).

Implications and Policy Recommendations

What can be done about this reality, in which privatized monopoly utilities effectively capture their regulators, advancing shareholders' interest at ratepayers'—and sometimes taxpayers'—expense? If there were an easy fix, it would likely have been implemented long ago. Yet there are some "not easy" fixes that are still worth discussing here.

First, as Sandeep Vaheesan has recently advocated, we could turn over all ownership of the electricity grid to the federal government (Vaheesan 2024a), which could either run the system directly itself, or Congress could "authorize states, cities, towns, and counties, individually or collectively, to take over investor-owned utilities serving them and operate them as democratic enterprises committed to full decarbonization" (Vaheesan 2024, 259). Vaheesan points out that publicly owned utilities already deliver power to significant parts of the country, for example, municipal utilities like the Los Angeles Department of Water and Power (LADWP). He also notes that the New Deal saw the creation of federal power agencies such as the Tennessee Valley Authority (TVA) and the Bonneville Power Administration (BPA), which generate electric power and are unencumbered by the imperative to deliver shareholder profits, allowing the *potential* for better responsiveness to the public (Vaheesan 2024). Eliminating IOUs would certainly be one way to rein in the obscene amount of corporate political power they wield, but the financial and legal complexity that this would create cannot be ignored and should not be understated.

There have been efforts in the past to municipalize investor-owned utilities, and they typically fail once people become aware of the up-front cost and legal challenges that would ensue. For example, in San Diego, organizers of a campaign to put a question on the ballot about a municipal takeover of San Diego Gas and Electric (SDG&E) failed to gather the requisite number of signatures (The Editorial Board 2024). In Chicago, similarly, plans for a municipal utility were abandoned when the up-front cost of a buy-out of ComEd was determined to be too high (Gheorghiu 2020). And in Maine, a *statewide* ballot measure to replace *both* of that state's major IOUs, Central Maine Power and Versant, with a consumer-owned utility to be named Pine Tree Power was defeated by a 2–1 margin after a campaign in which the side representing the status quo (i.e., retaining the IOUs) argued not only that the cost of the buyout for taxpayers would be too high to bear but also that costly and prolonged litigation would ensue, preventing the intended outcome (De Socio 2023). Indeed, these failures could be interpreted as even more evidence of the deeply structural power held by IOUs. Of course, what Vaheesan proposes is different and would be unprecedented: The *federal* government, with its greater powers including eminent domain, would be directly involved (Vaheesan 2024).

Legally, this might be easier, but politically, it would likely be even more challenging, requiring an act of Congress that not only would likely draw opposition (and perhaps also legal challenges) from states' rights proponents but would also unify the entire IOU industry against it. This industry is often divided regionally, and on a company-by-company basis, when it comes to policy minutia. By contrast, a grand proposal like this would likely be viewed as an existential threat that would foster immediate cooperation and unity within the industry. The industry is represented in Washington, DC, by the Edison Electric Institute, a powerful trade association, among others. These obstacles do not mean that national grid ownership would not be worth pursuing, but they do mean such an effort cannot necessarily be counted on to work.

From a climate perspective, it is also worth noting that a national public utility would also be no guarantee of a greener grid. Indeed, existing publicly owned utilities have a mixed record in terms of their resource portfolios, and it is not clear that publicly owned necessarily means greener. Consider, for example, the federally owned TVA. In 2023, 57 percent of the electricity TVA generated came from fossil-fueled sources, coal (18 percent) and gas (39 percent; Eggers 2024). By contrast, Xcel Energy—a vertically integrated IOU operating in eight states in the Western and Midwestern regions of the country—reported only 50 percent of the 2023 electricity it generated came from fossil-fueled sources (Xcel Energy 2024). Furthermore, despite being federal government–owned, there are signs of governance/coordination problems that render TVA recalcitrant to the influence of other forces within the administrative state. This was most plainly evident in a recently reported incident in which TVA skirted efforts on the part of the Biden Environmental Protection Agency (EPA) to subject its resource planning to further environmental review (Anchondo 2024). Of course, these could be addressed through additional public policies, but that requires additional mobilization and political effort.

In the absence of a federal government takeover, in many ways the best solution from the perspective of keeping IOUs' political power in check may be doing throughout the country what has already been done in the fully restructured states examined in this book. In addition to neutralizing the opposition of "transmission and distribution only" utilities to renewable energy mandates, as we have seen in chapters 3 and 4 of this book, there is an added benefit of electric utility sector restructuring from a corporate political power perspective: democratization of stakeholder-based policy-making processes.

More competition in the generation and/or retail segments of the utility sector translates into more interest group actors (such as IPPs and RPMs) lobbying the state legislature and intervening in PUC deliberations. While it does not guarantee that the *quality* of their representation will rival those of monopoly transmission and distribution utilities, it does stand to reason that legislators and regulators will have more (and different) perspectives to consider in their decision-making. And while it does not correct the overrepresentation of *private* (corporate) actors relative to *public interest* actors, the political science theory of interest group pluralism (e.g., Truman 1951) tells us that a divided business community should be better able to keep the influence of any single private actor, such as IOUs, in check. State-level restructuring alone does not solve the problem of corporate influence given the "multilayer subsidiary" structure of the current IOU industry (Prechel 2020), and the "hidden value transfers" that result (Kovvali and Macey 2022). But it does at least democratize the *state-level* policy landscape, at least to a degree.

Restructuring—the policy I am advocating here strictly from the perspective of reducing the state-level corporate political power of IOUs—is not without its critics. Criticism comes from at least two directions. First, there are those concerned about ensuring the reliability of the grid, who argue that a vertically integrated system has fewer uncertainties, fewer players, and thus is better equipped to ensure overall system reliability (Angwin 2020). Second, there is the argument that wholesale electricity markets (ISOs/RTOs), which go together with restructuring, lack adequate oversight by the FERC and therefore can result in their own form of corporate capture (Macey and van Emmerick 2025; Peskoe 2023; Vaheesan 2024). Evaluating these arguments in full is beyond the scope of this book, though the second one is especially compelling and points to the need for federal policy reform to accompany state-level restructuring.

Perhaps the easiest option when it comes to reducing IOUs' *state-level* influence is to support efforts, which have already been adopted in a few states, to use existing state authorities to limit IOUs' ability to pass the costs of their *political* work onto ratepayers (believe it or not, this is legal in most states). Utility watchdog groups like the Energy and Policy Institute have successfully passed legislation in states like Connecticut and

Colorado—meaning restructured and vertically integrated states alike—and continue to push for such legislation elsewhere (Energy and Policy Institute 2025). These policies are extremely necessary but hardly sufficient to prevent the type of influence documented in this book.

Another, more feasible (albeit partial) fix for some of the issues raised in this book is greater nonutility involvement in PUC decision-making. Ideally, this involvement would include not only more private sector actors but public interest groups as well. This is precisely the mission of a new organization called Powerlines, led by a former US Department of Energy official. Powerlines aims to raise public awareness about the importance of PUCs in the US energy landscape, and ultimately, to pursue reforms that modernize and democratize it (Roberts 2024).

Contributions to the Literature on US Climate and Renewable Energy Policymaking

This book benefits from a large existing literature on the politics of US climate and renewable energy policy, which emphasizes federalism and the critical role of state governments (e.g., Basseches et al. 2022; Bergquist and Warshaw 2023; Rabe 2013). This section details three key contributions that this book makes to that literature. First, it suggests the usefulness of moving beyond partisanship and energy economy as the most important determinants of policy prospects. To be clear, this does not mean partisanship and energy economy are not important nor that we should not continue to study their roles; they certainly explain some things, just not everything. Second, my findings suggest this literature should pay greater attention to IOUs, as well as questions of electricity grid ownership and governance. Third, in focusing on IOUs, there should be a greater appreciation of the variation that exists when it comes to their policy preferences regarding renewable energy. I discuss each of these in turn.

Limits in the Explanatory Power of Partisanship and Energy Economy

In chapters 3 and 4, we saw that political partisanship and energy economy—two dominant explanations in the literature for the viability of state-level climate policymaking (e.g. Bromley-Trujillo et al. 2016; Coley and Hess 2012; Trachtman 2020)—are circumscribed in their impact on RPS policy design. To be sure, these things do matter. For example, the timing and content of the RPS policy in Oregon was highly dependent on that state's Democratic Party–led government, whereas the limitations of South Carolina's purely voluntary RPS policy were largely attributable to that state's Republican-led government. Similarly, the limitations of Oklahoma's purely voluntary RPS policy can be largely attributed to that state's fossil fuel production–based economy. And yet, these factors were overcome in Texas and Pennsylvania, resulting in mandatory RPS policies that moved the needle, especially in Texas (which now leads the nation in wind energy and rivals California in solar) but also to a lesser extent in Pennsylvania (Wiser and Langniss 2001), due to those states having fully restructured their electric utility sectors.

The legislation that fully restructured those two states' utility sectors, thus neutralizing the opposition of certain IOUs to the RPS, in turn, was not a partisan exercise. The Texas restructuring law was passed by a Democrat-controlled House but a Republican-controlled Senate, and was signed by a Republican governor, George W. Bush (not typically thought of as a climate champion). In Pennsylvania, the restructuring law was passed by both a Republican-controlled House and a Republican-controlled Senate and was signed into law by a Republican governor, Tom Ridge. In both cases, the roll call votes were not split along party lines, and neither were the talking points that legislators delivered on the House or Senate floor (Pennsylvania House Journal 1996; Pennsylvania Senate Journal 1996; Texas State Archives 1999). Restructuring, whose feedback effects facilitated subsequent passage of the RPS, was not about partisan politics; it was about economics.

These findings join other sources of hope that the clean energy transition, at least in certain aspects, can be a more bipartisan or nonpartisan endeavor than previously thought. These other sources of hope include the economic benefits from the Inflation Reduction Act, which have flowed disproportionately to "red states" (Egan 2024), as well as the political geography of renewable energy production (Egan and Mullin 2023).[5] As this book has shown, the fundamentally political-economic nature of electricity policy, in which technical distinctions such as utility sector structure can have such a significant impact on interest groups' policy preferences on renewable energy legislation, should offer further hope that many of the changes that might facilitate clean energy policy in the future need not be about Democrats or Republicans, red states or blue states, but rather dollars and cents. A recent study has shown that partisanship was even less of a factor when it came to state-level tax policies incentivizing clean energy (Gilmore and St. Clair 2025).

At the same time, this reality is something of a double-edged sword. As Spence (2024) points out, intense grassroots polarization—which has been made far worse by the decline of traditional media and the rise of social media, and which has certainly gotten worse since the period that much of this book has focused on—threatens policymakers' ability to focus on these important technical matters in the future, in productive ways. We need the public to become more educated in such matters so that they can be influenced by reality rather than perception, which is increasingly biased by affective partisan polarization[6]; this is no easy task given the larger social forces at play (Spence 2024).

When it comes to in-state energy economy, we have seen that fossil fuel states like Texas and Pennsylvania are still capable of adopting at least moderately ambitious renewable energy policy. There are limits, to be sure. In the case of Texas, as Stokes (2020) documents, the initial momentum of a wind-focused RPS was blunted and completely halted when advocates in that state attempted to pass pro-solar policies several years later. And in Pennsylvania, a mandatory RPS was only possible due to major, undermining compromises on resource eligibility, such as allowing waste coal to count—resulting in that state's law being named the "*alternative* energy portfolio standard" (Glenna and Thomas 2010). Moreover, while it may be the case that high levels of in-state fossil fuel production do not necessarily preclude the passage of mandatory RPS policies, as Gazmararian and Tingley (2023) show, policies focused on a "just transition," such as those that promote clean energy investment or jobs targeted especially at communities that have historically been major fossil fuel producers, face unique credibility challenges in such communities.

Importance of IOUs, Grid Ownership, and Grid Governance

A second significant contribution of this book is that IOUs are critically important—and extremely politically powerful—actors in state-level renewable electricity policymaking, and by extension, climate policymaking more generally, given the decarbonization strategy of "electrify everything" (Fiorino 2022). Moreover, the degree to which IOUs own the grid, and are given the power to govern it (via fragmented governance structures and regulatory capture), matters enormously. Specifically, the literature would benefit from moving from thinking about the electricity grid exclusively or primarily in terms of the *resources* or *technologies* that power it to also thinking about it in terms of *who owns* and *who governs* it, and the implications in terms of energy sector actors' political and economic motivations and policy preferences.

When it comes to the question of public versus private ownership, it is tempting to divorce resources/technologies from questions of ownership because we see a range of "greenness" in the resource portfolios of both investor-owned and publicly owned utilities (Eggers 2024; Xcel Energy 2024). However, there are other factors besides the current resource portfolios to consider when thinking about what will be needed for a green grid in the *future*. In the future, now that costs of renewable energy have fallen considerably from the time the RPS policies studied in this book were first adopted, a major limiting factor on policymakers' willingness to promote green energy is likely to be concerns about reliability. Since wind and solar are intermittent resources, to what extent do we still need fossil fuel plants for backup power to keep the lights on? The importance of this question is magnified by the artificial intelligence (AI) revolution and the dramatic increase in electricity-powered data centers that is expected to result.

In a recent paper, legal scholars Klass, Macey, Welton, and Wiseman argue that "grid reliability governance suffers from excessive privatization" (Klass et al. 2024, 5). In other words, those we rely on to supply and interpret the very data upon which governance decisions related to reliability are made are typically private actors (IOUs or their corporate relatives), often with a financial self-interest in maintaining the status quo. How can we expect such actors' statements about the grid's reliability needs to not be biased against new, nonincumbent resources, many of which are lower or non-carbon-emitting—for example, the role that could be played by non-utility-owned, distributed energy resources such as rooftop solar, smart meters, or virtual power plants (Klass et al. 2024)?

Relatedly, achieving a green grid—regardless of the degree to which assets are publicly or privately owned—will require increasingly centralized planning, coordination, and governance. We have seen this already when it comes to the need to plan for, build, and fairly distribute the costs of interregional transmission, something that many believe requires an act of Congress to succeed but in the meantime has been facilitated by administrative actions such as FERC's Order 1920. The political power of incumbent

IOUs, who benefit financially from the status quo of decentralized grid governance, is an obstacle to this imperative for a reliable and affordable green grid (Hausman 2024).

Although these governance and centralization/decentralization issues are somewhat separate from the state-level analysis presented in this book, the implications of the stranglehold that IOUs have over the current system, especially in traditionally regulated, nonrestructured states, is an argument for pursuing state-level policy aimed at reducing their dominance, in concert with a transfer of powers from captured state PUCs to the federal government. The federal government is better positioned to consider cost allocation issues that will increasingly defy the political borders of individual states and is more resistant to capture by corporate utility structures whose geographic footprint also transcends state borders. In other words, while the structure of a state's utility sector is naturally more relevant to the distribution of interest group political power *in state capitals*, the lessons we can learn from studying state-level policymaking can be applied to bolster arguments for more centralized governance that can gain as much independence as possible from IOUs.

IOUs Are Always Powerful, but Their Preferences Can Vary

In the meantime, so long as we are stuck with the current system in which states govern many aspects of renewable energy policy, we would be well served to not assume IOUs are simply an extension of the fossil fuel industry, aligned in their policy preferences to discourage state-level renewable energy policy. My research finds that, in fully restructured states such as California, Massachusetts, Pennsylvania, and Texas, IOUs are divided over their renewable energy policy preferences. In these states, regulated transmission and distribution utilities are agnostic regarding the environmental stringency of RPS policies, caring only about their distributive impacts (Basseches 2024).

This does not mean that these companies do not pose obstacles to other climate policies, besides RPS policies. Indeed, the policies they are most likely to oppose are those that infringe upon their monopoly ownership of the transmission and distribution systems, such as solar net metering for distributed (as opposed to utility-scale) renewable energy resources. Indeed, in California, one of the states in which I have shown elsewhere that transmission and distribution IOUs' *support* of economy-wide GHG reduction and RPS policies was critical to their success (Basseches 2024), these same companies were largely responsible for a recent PUC decision to roll back net metering payments for customers who generate their own solar electricity (Gearino 2023).

All of this suggests the need for a deeper and more nuanced understanding of IOUs' policy preferences on specific climate and renewable energy policy proposals, rather than presuming their automatic opposition, and villainizing them accordingly. Although the research presented in this book directly challenges Spence's (2024) assertion that climate advocates and activists are too quick to invoke regulatory capture as the explanation for policy shortcomings, I do agree with Spence that "the claim that IOUs have impeded the transition is an imprecise generalization" (Spence 2024, 181). Instead of villainizing IOUs, we need to *study them* to better understand their motivations for the policy positions that they take.

Future Research Directions

This book has limitations that point to the need for future research on some of the issues it brings to the fore. The first is that while it examines states that vary in terms of partisanship, energy economy, and utility sector structure, it is less useful for understanding the politics of vertically integrated, nonrestructured states that have still managed to adopt stringent, mandatory RPS policies. The case of Oregon shows how Democratic partisanship combined with no in-state fossil fuel production can create an environment in which IOU resistance can be overcome and a relatively stringent RPS policy can still be adopted, but there are other vertically integrated states not among the seven investigated here that have still managed to adopt moderate to strong RPS policies (Carley et al. 2018). Future research could investigate such states.

Second, some RPS policies were adopted by ballot initiative rather than through the legislature. Colorado and Washington state are prominent examples. Although one quantitative study investigates determinants of county-level voting results for such measures (Pritchard and Mills 2021), this is an area in which there has been relatively little research. In particular, readers of this book may want to know what role, if any, IOU lobbying or utility sector structure plays behind the scenes in shaping such election outcomes. For example, to what extent do IOUs contribute to campaigns associated with either side of the ballot measure, and to what degree is their mobilization informed by similar or different considerations than their lobbying, as documented in this book in states with legislative adoption? Such questions would be ripe for investigation.

Finally, to what extent is utility sector structure, and the policy preferences of IOUs, relevant when considering other state-level climate and renewable energy policies besides RPS? It seems clear that IOUs are more likely to oppose net metering policies, regardless of utility sector structure (Stokes 2020). What about transportation or industrial electrification policies, tax policies, or energy efficiency programs and public benefit funds—to name just a few? All of these questions are ripe for investigation, and such investigations can consider insights from the research presented in this book.

Abbreviations

AARP	American Association of Retired Persons
ACP	Alternative compliance payment
AECT	Association of Electric Companies of Texas
AEPS	Alternative Energy Portfolio Standard
AI	Artificial intelligence
AIM	Associated Industries of Massachusetts
ALEC	American Legislative Exchange Council
ARIPPA	Appalachian Region Independent Power Producers Association
BPA	Bonneville Power Administration
Btu	British thermal units
CAISO	California Independent System Operator
CalChamber	California Chamber of Commerce
CalSEIA	California Solar Energy Industries Association
CalWEA	California Wind Energy Association
CCL	Coastal Conservation League
CCS	Carbon capture and sequestration
CEERT	Center for Energy Efficiency and Renewable Technologies
ComEd	Commonwealth Edison
ConEd	Consolidated Edison
CTC	Competitive transition charge
CUB	Citizens Utility Board
CVSC	Conservation Voters of South Carolina
CWIP	Construction work in progress
DER	Distributed energy resource
DG	Distributed generation

DSM	Demand-side management
EAC	Energy Advisory Council
EDF	Environmental Defense Fund
EPA	Environmental Protection Agency
ERCOT	Electric Reliability Council of Texas
FERC	Federal Energy Regulatory Commission
FPA	Federal Power Act of 1920
FP&L	Florida Power and Light
GHG	Greenhouse gas
IECPA	Industrial Energy Consumers of Pennsylvania
IOU	Investor-owned utility
IPP	Independent power producers
IRA	Inflation Reduction Act of 2022
IRP	Integrated resource plan
ISO	Independent system operator
ISO-NE	New England Independent System Operator
ITC	Intangible transition charge
LADWP	Los Angeles Department of Water and Power
MDOER	Massachusetts Department of Energy Resources
Met-Ed	Metropolitan Edison
MISO	Midcontinent Independent System Operator
MWh	Megawatt hour
NCSL	National Conference of State Legislatures
NECEC	New England Clean Energy Council
NEM	Net energy metering
NRDC	Natural Resources Defense Council
NWFPA	Northwest Food Processors Association
OCA	Office of the Consumer Advocate (Pennsylvania)
OCC	Oklahoma Corporation Commission
OEC	Oregon Environmental Council
OG&E	Oklahoma Gas and Electric
OLCV	Oregon League of Conservation Voters
OSBA	Office of the Small Business Advocate (Pennsylvania)
PDEP	Pennsylvania Department of Environmental Protection
PECO	Philadelphia Electric Company
Penelec	Pennsylvania Electric
PennPower	Pennsylvania Power Company

PGE	Portland General Electric
PG&E	Pacific Gas and Electric
PJM	PJM Interconnection
POU	Publicly owned utility
PP	Pacific Power
PP&L	Pennsylvania Power and Light
PSO	Public Service Co. of Oklahoma
PTC	Production tax credit
PUC	Public utility commission
PUD	Public utility district
PUHCA	Public Utility Holding Company Act of 1935
PURPA	Public Utility Regulatory Policies Act of 1978
PV	Photovoltaic
QF	Qualifying facility
REC	Renewable energy credit
REP	Retail electric provider
RPM	Retail power marketer
RPS	Renewable portfolio standard
RTO	Regional transmission organization
SACE	Southern Alliance for Clean Energy
SCE	Southern California Edison
SCE&G	South Carolina Electric and Gas
SCPSC	South Carolina Public Service Commission
SDG&E	San Diego Gas and Electric
SEC	Securities and Exchange Commission
SELC	Southern Environmental Law Center
SPP	Southwest Power Pool
TAB	Texas Association of Business
TAM	Texas Association of Manufacturers
TIEC	Texas Industrial Energy Consumers
TREIA	Texas Renewable Energy Industries Association
TVA	Tennessee Valley Authority
UGI	UGI Utilities
WECC	Western Electricity Coordinating Council
WSPA	Western States Petroleum Association

Notes

Chapter 1

1. Today, the names of these bodies vary by state, but their essential function is the same.

2. It is transmission lines of this sort that many are calling for to realize the full decarbonization potential of the federal Inflation Reduction Act, and this underlying need for transmission expansion underlies clean energy advocates' push for federal permitting reform to speed up the construction of these interstate transmission lines (Jenkins et al. 2022).

3. For readers wondering why Texas has its own grid, read on.

4. Electricity customers are also known as "ratepayers." Henceforth, the two terms will be used interchangeably.

5. "Distribution" also encompasses what is known as distributed generation (DG) or distributed energy resources (DERs). Ownership of these resources and the wires connecting them to the rest of the grid is hotly contested between utility companies and customers, resolved only in part by net energy metering (NEM) policies. DG and DERs are increasingly prevalent as the clean energy transition progresses, and include rooftop solar panels, distributed wind, battery storage, and other forms of electric generation and storage that are sited with the customer as opposed to utility-scale generation, which includes traditional nuclear, gas-fired, and coal-fired power plants as well as large wind farms and solar arrays, among other technologies.

6. Although, in the case of the California Independent System Operator (CAISO), board members are appointed by the California governor, which arguably creates slightly more accountability to the public than elsewhere (Vaheesan 2024).

7. Prior to being FERC, this agency was known as the Federal Power Commission (Boyd 2013).

8. Perhaps federal officials were compelled to act because, in 1930, just nineteen holding companies controlled 90 percent of IOUs (Isser 2015).

9. Although, many IPPs were subsequently bought out by giant energy holding companies, which owned both IOUs and IPP affiliates, raising questions about how independent IPPs really were (see Hempling 2020, Kovvali and Macey 2022).

10. It is important to note that the balance of non-IPP-owned generation is not *necessarily* owned by IOUs (though across the fifty states, in aggregate, most of it is). It could also be owned by POUs or self-generators, such as rooftop solar owners, which explains why, for example, California IPPs generated only about 54 percent of the state's electricity in 2021 despite IOUs having been required to divest their generation in that state as part of its restructuring law.

11. See Vaheesan (2024) for more on POUs and their politics.

Chapter 2

1. The American Clean Energy Security Act, better known as Waxman–Markey, passed the US House of Representatives but died in the US Senate.

2. They were both adopted in the same state, during the same legislative session.

3. Percentage increase of target divided by number of years to meet it, multiplied by the percentage of electricity load to which the mandate applies

4. I am grateful to Sanya Carley for sharing her raw data, which assigns California a stringency score 4.3468, Massachusetts a stringency score of 2.77103, Oregon a stringency score of 3.25823, Pennsylvania a stringency score of 2.7428, Texas a stringency score of 2.63966, Oklahoma a stringency score of 0.810967, and South Carolina a stringency score of –0.961861.

5. California Senate Bill 1078, amending Public Utility Code §399.15, chapter 516, California Acts of 2002.

6. California Senate Bill 107, amending Public Utility Code §399.14, chapter 464, California Acts of 2006.

7. California Senate Bill 2X, amending Public Utility Code §399.15, chapter 1, Special Session Acts of 2011.

8. Massachusetts Session Laws, Acts of 1997, chapter 164.

9. 225 CMR 14.00.

10. Massachusetts Senate Bill 2768, amending §1A of MGL, chapter 164, Session Law Acts of 2008.

11. Oregon Senate Bill 838, C-Engrossed, Session Acts of 2007.

12. Texas Senate Bill 7, §27, Texas General Session Laws of 1999.

13. Texas Senate Bill 20a, §1, Texas Special Session Laws of 2005.

14. Pennsylvania Senate Bill 1030, §2, 2004 Laws of the General Assembly of the Commonwealth of Pennsylvania.

15. Pennsylvania Senate Bill 1030, §3.

16. As table 2.5 indicates, state record-keeping vigilance varied considerably from one state to the next, and along with it the extent and quality of the information available in the state archives.

Chapter 3

1. This includes the Western US states of Montana, Wyoming, Colorado, New Mexico, Arizona, Utah, Idaho, Nevada, California, Oregon, and Washington as well as the Western Canadian provinces of Alberta and British Columbia.

2. A pseudonym for an environmental group lobbyist, whose real name is not being used to maintain anonymity.

3. The 2019 law that updated the South Carolina's solar net metering policy, among other things.

4. Formerly South Carolina Electric and Gas.

5. Two Duke Energy subsidiaries operate in South Carolina. Duke Energy Carolinas serves the lion's share of Duke customers in the state, but about 10–11 percent are served by Duke Energy Progress, headquartered in North Carolina.

6. Decoupling is part of a suite of policies known as "demand-side management," whose goal is to reduce electricity demand. It divorces revenue from volume of sales in utility ratemaking processes.

7. Several people I interviewed believe the decision to rebrand "South Carolina Electric and Gas" as "Dominion" was an effort to salvage the company's reputation after taking such a big hit due to the VC Summer scandal.

8. A public service commission (PSC) is the functional equivalent of a PUC. While these agencies are most often called PUCs, in some states they are called PSCs, and in a few states, they have other names (e.g., the State Corporation Commission in Virginia, the Commerce Commission in Illinois, the Public Utility Regulatory Authority in Connecticut, and the Department of Public Utilities in Massachusetts).

9. In Oregon, the majority party leader—in this case, the Speaker of the House—appoints committee chairs who are copartisans.

10. An organizational sponsor of The Heritage Foundation's *Project 2025*.

11. Interestingly, one Texas PUC regulator and prominent environmental advocate both recall Bush telling them, "Oh, hey. By the way, we like wind" (author's

interviews, 3/11/22, 3/18/22). "He would sometimes talk in the plural," the regulator explained (author's interview, 3/18/22).

12. SB 533 was the regular session bill whose language was identical to SB 20, which ultimately passed during a special session the same year.

13. It is worth nothing, however, that in the non-ERCOT regions of Texas, IOUs remained vertically integrated and opposed to the RPS, as evidenced by testimony provided by Entergy, a vertically integrated IOU operating in the Gulf Coast region (Public Utility Commission of Texas 1999).

Chapter 4

1. Oklahoma City's NBA basketball team.

2. A pseudonym.

3. A pseudonym.

4. A pseudonym.

5. The client's name is omitted to protect the anonymity of the interviewee.

6. One could now add many others to this list, including but not limited to New York, the other New England states, and even Midwestern leaders like Minnesota, Illinois, or, most recently, Michigan.

7. A national Democratic strategist who hails from the Deep South.

8. Significant nuclear generation accounted for the vast majority of the rest.

9. At the time the Pennsylvania RPS policy was being proposed, many within and outside the industry saw natural gas as a "bridge fuel" to ease the transition from coal to renewables, and undoubtedly the gas industry—which after all did need to compete with the coal industry—recognized certain business opportunities associated with reforming the state's energy policy (author's interviews, multiple).

10. A pseudonym.

11. Industrial ratepayers enjoyed high-quality representation in all seven states in this book.

12. It is common in other states for attorneys general offices themselves to perform ratepayer advocacy functions, but Pennsylvania is distinct from other states in this book in having separate, dedicated state agencies for this work.

13. A group active in many of the other states in this book as well.

14. This company was previously owned by Allegheny Energy but was acquired by First Energy in 2010 (Joint Press Release of Allegheny and First Energy 2010).

15. This bill number was initially introduced by Republican Representative Kathrynann Durham as legislation dealing with taxicab medallions (a subject also regulated by the state PUC), but was gutted and replaced with the agreed-upon language related to electricity restructuring in a backroom deal that took place while the bill was in committee (author's interviews, multiple).

16. Including a defeated amendment by Democratic State Senator Allyson Schwartz, which would have introduced an RPS requiring 5 percent renewable electricity by 2007.

17. Despite State Senator Allyson Schwartz's attempt to include one by way of a floor amendment (Pennsylvania Senate Journal 1996).

18. Known as "power purchase agreements" (PPAs)

19. The term, "qualified facilities" (QFs), comes from PURPA and refers to small generators that are given special status under federal law and which regulated electric utilities are encouraged (and in some cases required) to contract with for power.

20. "Retail wheeling" was a term that was used interchangeably with "retail choice" and "retail competition" at the time.

21. Universal service funds are intended to ensure the provision of electricity to all customers ("universally") regardless of whether they take advantage of the competitive retail market and including certain protections for low-income ratepayers and energy conservation funds are intended to incentivize energy efficiency. These are both examples of what are known as "public benefit funds" (Prasad 2023) as were the aforementioned funds to encourage renewable energy in utilities' service territories.

22. A provision important to organized labor.

23. See Basseches 2024 for similar examples from other states.

24. The distinction between tier 1 and tier 2 resources will soon be discussed.

25. Tier 2 also included demand-side management, as an incentive for utilities to reduce demand.

26. An RTO that includes Pennsylvania, New Jersey, Maryland, Delaware, Virginia, West Virginia, Ohio, and small parts of Kentucky and North Carolina.

27. A pseudonym.

28. A "corporate relative" of PECO.

29. A "corporate relative" of Penn Power, Met-Ed, Penelec, and West Penn Power, after 2010 when FirstEnergy acquired Allegheny Energy.

30. A "corporate relative" of Pennsylvania Power and Light.

Chapter 5

1. E.g., Washington, Montana, Colorado Arizona, Hawaii, New Mexico, North Carolina.

2. Previously South Carolina Electric and Gas/SCANA.

3. This is in contrast to intermittent electricity provided by weather-dependent resources such as wind and solar. The US Energy Information Administration defines "base load plants" as those that are "normally operated to take all or part of the minimum load of a system, and that consequently produce electricity at an essentially constant rate and run continuously." In the United States, such plants are typically coal-fired, gas-fired, nuclear, and sometimes hydroelectric.

4. A pseudonym.

5. The state capital.

6. See chapter 4 for more on the reasons their effort ultimately failed.

7. A pseudonym.

8. Massachusetts adopted its mandatory RPS in 1997; Texas did so in 1999; California did so in 2002; Pennsylvania did so in 2004.

9. This refers to the ability of the utility to obtain special low-interest financing due to the guarantee ("security") of being able to repay the loans using ratepayer money.

10. The repeal was part of HB 1090, which passed in 2007.

11. Note that the "just and reasonable" standard is less stringent (more utility-friendly) than the "used and useful" standard.

12. As the previous chapter describes, these cases were thousands of pages long, many of which were redacted, and despite my best efforts to obtain those pages that were made public, the Pennsylvania PUC's web portal is extremely poorly organized and does not allow me to confidently interpret the cases' content.

13. Analogous to "renewable energy credits" in other states.

Chapter 6

1. For instance, the FirstEnergy scandal in Ohio, in which the utility parent company paid more than $60 million in bribes to the former Speaker of the House and chairman of the Public Utilities Commission (Schladen and Trau 2024).

2. Arguably even more so given how often the legislature delegates the details to the rulemakings at the PUC.

3. This may be even more true of appointed commissioners as opposed to electeds, since their appointments could be made with their level of prior relevant experience in mind.

4. See Basseches 2024 on "flexible forms of corporate organization" as a source of IOUs' power.

5. Although see Oatley (2024) for challenges associated with political geography.

6. For more on the concept of "affective partisan polarization," see Druckman et al. 2018.

References

Abramowitz, Alan I. 2022. "The Polarized American Electorate: The Rise of Partisan-Ideological Consistency and its Consequences." *Political Science Quarterly* 137 (4): 645–674.

Aklin, Michael, and Matto Mildenberger. 2020. "Prisoners of the Wrong Dilemma: Why Distributive Conflict, Not Collective Action, Characterizes the Politics of Climate Change." *Global Environmental Politics* 20 (4): 4–27.

American Oil and Gas Historical Society. 2024. "Making Tulsa 'Oil Capital of the World.'" Accessed March 26, 2024. https://aoghs.org/petroleum-pioneers/glen-pool-tulsa-oil-capital/.

Anchondo, Carlos. 2024. "EPA Balked at TVA's Fossil Fuel Plant, Then Stepped Aside." *E&E News*, June 5.

Anderson, Taylor W. 2016. "Transcript of Bulletin Interview with Gov. Kate Brown." *The Bulletin*, April 10.

Angwin, Meredith. 2020. *Shorting the Grid: The Hidden Fragility of Our Electric Grid.* Wilder, VT: Carnot Communications.

Ansolabehere, Stephen, and David M. Konisky. 2016. *Cheap and Clean: How Americans Think About Energy in the Age of Global Warming.* Cambridge, MA: MIT Press.

Anzia, Sarah F. 2019. "Looking for Influence in All the Wrong Places." *Journal of Politics* 81 (1): 343–351.

Anzia, Sarah F., and Molly C. Jackman. 2013. "Legislative Organization and the Second Face of Power: Evidence from US State Legislatures." *Journal of Politics* 75 (1): 201–224.

Bade, Gavin. 2019. "San Francisco Considering Municipal Utility as PG&E Prepares Bankruptcy Filing." *Utility Dive*, January 15.

Barbose, Galen L. 2021. "US Renewables Portfolio Standards 2021 Status Update: Early Release." Lawrence Berkeley National Lab. Berkeley, CA.

Barbose, Galen L. 2023. "US Renewables Portfolio Standards 2023 Status Update." Lawrence Berkeley National Lab. Berkeley, CA.

Basseches, Joshua A. 2019. "'It Happened Behind Closed Doors:' Legislative Buffering as an Informal Mechanism of Political Mediation." *Mobilization: An International Quarterly* 24 (3): 365–388.

Basseches, Joshua A. 2024. "Who Pays for Environmental Policy? Business Power and the Design of State-Level Climate Policies." *Politics and Society* 52 (3): 409–451.

Basseches, Joshua A., Michael C. Campbell, and Heather Schoenfeld. 2024. "Leveraging the Insights of Depth: A Staged Strategy for Building Qualitative Case Studies of American State-Level Policy." *Social Science Quarterly* 105 (2): 359–373.

Basseches, Joshua A., Rebecca Bromley-Trujillo, Maxwell T. Boykoff, Trevor Culhane, Galen Hall, Noel Healy, David J. Hess, David Hsu, Rachel M. Krause, Harland Prechel, J. Timmons Roberts, and Jennie C. Stephens. 2022. "Climate Policy Conflict in the US States: A Critical Review and Way Forward." *Climatic Change* 170 (3): 32.

Baumgartner, Frank R., Jeffrey M. Berry, Marie Hojnacki, David C. Kimball, and Beth L. Leech. 2009. *Lobbying and Policy Change: Who Wins, Who Loses, and Why?* Chicago: University of Chicago Press.

Bawn, Kathleen, Martin Cohen, David Karol, Seth Masket, Hans Noel, and John Zaller. 2012. "A Theory of Political Parties: Groups, Policy Demands and Nominations in American Politics." *Perspectives on Politics* 10 (3): 571–597.

Bergquist, Parrish, and Christopher Warshaw. 2023. "How Climate Policy Commitments Influence Energy Systems and the Economies of US States." *Nature Communications*.

Bergquist, Parrish, Matto Mildenberger, and Leah C. Stokes. 2020. "Combining Climate, Economic and Social Policy Builds Public Support for Climate Action in the US" *Environmental Research Letters* 15 (5): 054019.

Blunt, Katherine. 2022. *California Burning*. London: Penguin.

Borenstein, Severin, and James Bushnell. 2015. "The US Electricity Industry After 20 Years of Restructuring." *Annual Review of Economics* 7 (1): 437–463.

Boyd, William. 2013. "Public Utility and the Low-Carbon Future." *UCLA Law Review* 61: 1614.

Boyd, William, and Ann E. Carlson. 2016. "Accidents of Federalism: Ratemaking and Policy Innovation in Public Utility Law. *UCLA Law Review* 63: 810.

Bromley-Trujillo, Rebecca, J. S. Butler, John Poe, and Whitney Davis. 2016. "The Spreading of Innovation: State Adoptions of Climate Change Policy." *Review of Policy Research* 33 (5): 544–565.

Broockman, David E. 2012. "The 'Problem of Preferences:' Medicare and Business Support for the Welfare State." *Studies in American Political Development* 26 (2): 83–106.

Brulle, Robert J. 2013. "Institutionalizing Delay: Foundation Funding and the Creation of US Climate Change Counter-movement Organizations." *Climatic Change* 122 (4): 681–694.

Brulle, Robert J., and Christian Downie. 2022. "Following the Money: Trade Associations, Political Activity and Climate Change." *Climatic Change* 175 (3): 11.

Burke, Matthew J., and Jennie C. Stephens. 2017. "Energy Democracy: Goals and Policy Instruments for Sociotechnical Transitions." *Energy Research and Social Science* 33: 35–48.

Campbell, Andrea Louise. 2012. "Policy Makes Mass Politics." *Annual Review of Political Science* 15 (1): 333–351.

Carley, Sanya. 2011. "The Era of State Energy Policy Innovation: A Review of Policy Instruments." *Review of Policy Research* 28 (3): 265–294.

Carley, Sanya, and Chris J. Miller. 2012. "Regulatory Stringency and Policy Drivers: A Reassessment of Renewable Portfolio Standards." *Policy Studies Journal* 40 (4): 730–756.

Carley, Sanya, and David M. Konisky. 2020. "The Justice and Equity Implications of the Clean Energy Transition." *Nature Energy* 5 (8): 569–577.

Carley, Sanya, Lincoln L. Davies, David B. Spence, and Nikolaos Ziorgiannis. 2018. "Empirical Evaluation of the Stringency and Design of Renewable Portfolio Standards." *Nature Energy* 3 (9): 754–763.

Carley, Sanya, and David M. Konisky. 2020. *Power Lines: The Human Costs of American Energy in Transition.* Chicago: University of Chicago Press.

Carpenter, Daniel, and David A. Moss. 2014. *Preventing Regulatory Capture: Special Interest Influence and How to Limit It.* New York: Cambridge University Press.

Charmaz, Kathy. 2006. *Constructing Grounded Theory: A Practical Guide Through Qualitative Analysis.* Thousand Oaks, CA. Sage Publishing.

Chen, Anthony S. 2019. "In the Private Interest?" In *Can America Govern Itself?* edited by Frances E. Lee and Nolan McCarty. New York: Cambridge University Press.

Ciulla, Jessie, Gennelle Wilson, and Rachel Gold. 2022. "What Utility Regulators Need to Know about the Inflation Reduction Act." Rocky Mountain Institute.

Cohn, Julie. 2019. "When the Grid Was the Grid: The History of North America's Brief Coast-to-Coast Interconnected Machine." *Proceedings of the IEEE* 107 (1): 232–243.

Coley, Jonathan S., and David J. Hess. 2012. "Green Energy Laws and Republican Legislators in the United States." *Energy Policy* 48: 576–583.

Costello, Kenneth W., and Robert E. Burns. 2000. "Regional Transmission Organizations and the Coordination of Regional Electricity Markets: A Review of FERC Order 2000." *The National Regulatory Research Institute*. Columbus, OH: Ohio State University.

Cudahy, Richard D. 1998. "The Folklore of Deregulation (with apologies to Thurman Arnold)." *Yale Journal on Regulation* 15: 427.

Culhane, Trevor, Galen Hall, and J. Timmons Roberts. 2021. "Who Delays Climate Action? Interest Groups and Coalitions in State Legislative Struggles in the United States." *Energy Research and Social Science* 79: 102114 .

Cullenward, Danny, and David G. Victor. 2020. *Making Climate Policy Work*. Hoboken, NJ: Wiley.

Curry, James M. 2015. *Legislating in the Dark: Information and Power in the House of Representatives*. Chicago: University of Chicago Press.

De Socio, Mike. 2023. "What the Failed Pine Tree Power Proposal in Maine Could Have Accomplished." *CNET*, November 8.

Downie, Christian. 2019. *Business Battles in the US Energy Sector: Lessons for a Clean Energy Transition*. New York: Routledge.

Druckman, James N., Matthew S. Levendusky, and Audrey McLain. 2018. "No Need to Watch: How the Effects of Partisan Media Can Spread Via Interpersonal Discussions." *American Journal of Political Science* 62 (1): 99–112.

Drutman, Lee, Matt Grossmann, and Timothy LaPira. 2019. "The Interest Group Top Tier" In *Can America Govern Itself?* edited by Frances E. Lee and Nolan McCarty. New York: Cambridge University Press.

Egan, Matt. 2024. "Red States are Big Winners of Biden's Landmark Laws." *CNN.com*, February 14. https://www.cnn.com/2024/02/14/business/manufacturing-jobs-biden/index.html.

Egan, Patrick J., and Megan Mullin. 2023. "US Partisan Polarization on Climate Change: Can Stalemate Give Way to Opportunity?" *PS: Political Science and Politics*.

Eggers, Caroline. 2024. "TVA's Power Mix Explained in Two Charts." *WPLN*, April 12. https://wlpn.org/post/tvas-power-mix-explained-in-two-charts.

Energy and Policy Institute. 2025. "Tracking State Legislation to Get Politics Out of Utility Bills." Accessed April 23, 2025. https://energyandpolicy.org/tracking-states-getting-politics-out-of-utility-bills/.

Environmental Protection Agency. 2023. "Electricity Customers." Accessed July 23, 2023. https://www.epa.gov/energy/electricity-customers.

Exelon Corporation. 2023. "Our Companies." https://www.exeloncorp.com/companies. Accessed June 29, 2023.

Fiorino, Daniel J. 2022. *The Clean Energy Transition: Policies and Politics for a Zero-Carbon World.* New York: Polity.

Fischlein, Miriam, and Timothy M. Smith. 2013. "Revisiting Renewable Portfolio Standard Effectiveness: Policy Design and Outcome Specification Matter." *Policy Sciences* 46 (3): 277–310.

Ford, Lacy K., and Jared Bailey. 2022. *Empowering Communities: How Electric Cooperatives Transformed Rural South Carolina.* Columbia, SC: University of South Carolina Press.

Fortune Magazine. 2023. "Largest US Corporations." 187 (3). Accessed June 29, 2023.

Gazmararian, Alexander F., and Dustin Tingley. 2023. *Uncertain Futures: How to Unlock the Climate Impasse.* New York: Cambridge University Press.

Gearino, Dan. 2023. "California Enters 'Uncharted Territory' After Cutting Payments to Rooftop Solar Owners by 75 Percent." *Inside Climate News*, April 27.

George, Alexander L., and Andrew Bennett. 2005. *Case Studies and Theory Development in the Social Science.* Cambridge, MA: MIT Press.

Gheorghiu, Iulia. 2020. "Cutting Ties with ComEd Could Cost Chicago $8.8B, Report Finds." *Utility Dive*, September 1.

Gilens, Martin, and Benjamin I. Page. 2014. "Testing Theories of American Politics: Elites, Interest Groups, and Average Citizens." *Perspectives on Politics* 12 (3): 564–581.

Gilmore, Elisabeth, and Travis St. Clair. 2025. "Tax Expenditures as Tools for State-Level Climate Action in the US" *Climate Policy.*

Glenna, Leland L., and Robert Roy Thomas. 2010. "From Renewable to Alternative: Waste Coal, the Pennsylvania Alternative Energy Portfolio Standard, and Public Legitimacy." *Society and Natural Resources* 23 (9): 856–871.

Goodman, Sara Wallace. 2024. "Case Study and Selection." In *Doing Good Qualitative Research*, edited by Cyr and Goodman. New York: Oxford University Press.

Gormley, William T. 1983. *The Politics of Public Utility Regulation.* Pittsburgh, PA: University of Pittsburgh Press.

Grumbach, Jacob M. 2022. *Laboratories Against Democracy: How National Parties Transformed State Politics.* Princeton, NJ: Princeton University Press.

Hacker, Jacob S., Alexander Hertel-Fernandez, Paul Pierson, and Kathleen Thelan. 2021. "The American Political Economy: A Framework and Agenda for Research." In *The American Political Economy.* New York: Cambridge University Press.

Hacker, Jacob S., and Paul Pierson. 2014. "After the 'Master Theory': Downs, Schattschneider, and the Rebirth of Policy-Focused Analysis." *Perspectives on Politics* 12 (3): 643–662.

Hall, Galen, Trevor Culhane, and J. Timmons Roberts. 2024. "Climate Coalitions and Anti-Coalitions: Lobbying Across State Legislatures in the United States." *Energy Research and Social Science.*

Hall, Richard L., and Alan V. Deardorff. 2006. "Lobbying as Legislative Subsidy." *American Political Science Review* 100 (1): 69–84.

Harper, Charles, Sam Krasnow, Leah Stokes, Lissa Lynch, Sam Ricketts, Amanda Levin, Daniela Schulman, Jeff Slyfield, and Christy Walsh. 2023. "Powering Toward 100 Percent Clean Power by 2035: The Path to Carbon-Free Electricity After the Inflation Reduction Act." Evergreen Action and NRDC.

Harrison, Conor, and Shelley Welton. 2021. "The States That Opted Out: Politics, Power, and Exceptionalism in the Quest for Electricity Deregulation in the United States South." *Energy Research and Social Science* 79: 102147.

Haseloff, Thomas. 2024. "Who Shapes Environmental Policy? Power and Politics in the Renewable Portfolio Standards of US States, 1997–2020." *Socius* 10: 1–13.

Hausman, Catherine. 2024. "Power Flows: Transmission Lines, Allocative Efficiency, and Corporate Profits." *NBER Working Paper Series.* Working Paper 32091. http://www.nber.org/papers/w32091.

Heern, Jared. 2023. "Who's Controlling Our Energy Future? Industry and Environmental Representation on United States Public Utility Commissions." *Energy Research and Social Science.*

Hempling, Scott. 2020. *Regulating Mergers and Acquisitions of US Electric Utilities: Industry Concentration and Corporate Complication.* Northampton, MA: Elgar.

Henderson, Geoffrey, Alexander Hertel-Fernandez, Matto Mildenberger, and Leah C. Stokes. 2023. "Conducting the Heavenly Chorus: Constituent Contact and Provoked Petitioning in Congress." *Perspectives on Politics* 21 (1): 191–208.

Hertel-Fernandez, Alexander. 2019. *State Capture: How Conservative Activists, Big Businesses, and Wealthy Donors Reshaped the American States—and the Nation.* New York: Oxford University Press.

Hertel-Fernandez, Alexander, Matto Mildenberger, and Leah C. Stokes. 2019. "Legislative Staff and Representation in Congress." *American Political Science Review* 113 (1): 1–18.

Hirsh, Richard F. 1999. *Power Loss: The Origins of Deregulation and Restructuring in the American Electric Utility System.* Cambridge, MA: MIT Press.

Holdman, Jessica. 2023. "SC's Nuclear Fiasco: What's Next in Saga Surrounding Fraud and Mismanagement at VC Summer." *Post and Courier,* August 5, 2023.

Huang, Ming-Yuan, Janaki R. R. Alavalapati, Douglas R. Carter, and Matthew H. Langholtz. 2007. "Is the Choice of Renewable Portfolio Standards Random?" *Energy Policy* 35 (11): 5571–5575.

Hughes, Thomas Parke. 1983. *Networks of Power: Electrification in Western Society, 1880–1930.* Washington, DC: Johns Hopkins University Press.

Hurlbut, David. 2008. "A Look Behind the Texas Renewable Portfolio Standard: A Case Study." *Natural Resources Journal* 48 (1): 129–161.

Isser, Steve. 2015. *Electric Restructuring in the United States: Markets and Policy from the 1978 Energy Policy Act to Present.* New York: Cambridge University Press.

Jenkins, Jesse D., Jamil Farbes, Ryan Jones, Neha Patankar, and Greg Schivley. 2022. "Electricity Transmission is Key to Unlock the Full Potential of the Inflation Reduction Act." REPEAT Project. Princeton, NJ.

Jerolmack, Colin, and Edward T. Walker. 2018. "Please in My Backyard: Quiet Mobilization in Support of Fracking in an Appalachian Community." *American Journal of Sociology* 124 (2): 479–516.

Joint Press Release of Allegheny and First Energy. 2010. "First Energy and Allegheny Energy to Combine in $8.5 Billion Stock-For-Stock Transaction." February 11, 2010.

Journal Record Staff. 2023. "Bill Would Give Oklahoma Consumers Choice in Picking Power Provider." *The Journal Record,* February 14. https://journalrecord.com/2023/02/14/bill-would-give-oklahoma-consumers-choice-in-picking-power-provider/. Accessed July 7, 2023.

Kamieniecki, Sheldon. 2006. *Corporate America and Environmental Policy: How Often Does Business Get Its Way?* Stanford, CA: Stanford University Press.

Karapin, Roger. 2020. "Federalism as a Double-Edged Sword: The Slow Energy Transition in the United States." *Journal of Environment and Development* 29 (1): 26–50.

Kelsey, Nina, and Jonas Meckling. 2018. "Who Wins in Renewable Energy? Evidence from Europe and the United States." *Energy Research and Social Science* 37: 65–73.

Kinnard, Meg, and Jeffrey Collins. 2018. "Regulators OK $15B Merger of Troubled South Carolina Utility." *Associated Press State Wire: South Carolina,* December 14.

Kirkland, Justin H. 2014. "Ideological Heterogeneity and Legislative Polarization in the United States." *Political Research Quarterly* 67 (3): 533–546.

Klass, Alexandra, Joshua Macey, Shelley Welton, and Hannah Wiseman. 2024. "The Key to Electric Grid Reliability: Modernizing Governance." *Kleinman Center for Energy Policy.*

Kovvali, Aneil, and Joshua C. Macey. 2022. "Hidden Value Transfers in Public Utilities." *University of Pennsylvania Law Review* 171: 2129.

Kovvali, Aneil, and Joshua C. Macey. 2023. "The Corporate Governance of Public Utilities." *Yale Journal on Regulation* 40: 569.

Langniss, Ole, and Ryan Wiser. 2003. "The Renewables Portfolio Standard in Texas: An Early Assessment." *Energy Policy* 31 (6): 527–535.

LaPira, Timothy M., and Herschel F. Thomas. 2017. *Revolving Door Lobbying: Public Service, Private Influence, and the Unequal Representation of Interests*. Lawrence, KS: University of Kansas Press.

Layzer, Judith A. 2007. "Deep Freeze: How Business Has Shaped the Global Warming Debate in Congress." In *Business and Environmental Policy*, edited by Michael E. Kraft and Sheldon Kamieniecki. Cambridge, MA: MIT Press.

Lee, Nathan R. 2020. "When Competition Plays Clean: How Electricity Market Liberalization Facilitated State-Level Climate Policies in the United States." *Energy Policy* 139: 111308.

Lewin, Doug. 2025. "Texas' Energy Future: A Conversation with Jimmy Glotfelty." *The Texas Energy and Power Newsletter*, February 6, 2025.

Lupton, Robert N., Steven M. Smallpage, and Adam M. Enders. 2020. "Values and Political Predispositions in the Age of Polarization: Examining the Relationship between Partisanship and Ideology in the United States, 1988–2012." *British Journal of Political Science* 50 (1): 241–260.

Lupica, Lois R. 2000. "Transition Losses in the Electric Power Market: A Challenge to the Premises Underlying the Arguments for Compensation." *Rutgers Law Review* 52 (3): 649–718.

Lyon, Thomas P. 2016. "Drivers and Impacts of Renewable Portfolio Standards." *Annual Review of Resource Economics* 8: 141–155.

Lyon, Thomas P., and Haitao Yin. 2010. "Why Do States Adopt Renewable Portfolio Standards? An Empirical Investigation." *Energy Journal* 31 (3): 133–157.

Macey, Joshua, and Elias van Emmerick. 2025. "Towards a National Transmission Planning Authority." *Harvard Environmental Law Review* 49 (1): 79–147.

Matisoff, Daniel C. 2008. "The Adoption of State Climate Change Policies and Renewable Portfolio Standards: Regional Diffusion or Internal Determinants?" *Review of Policy Research* 25 (6): 527–546.

Maykuth, Andrew. 2012. "Pennsylvania's Acclaimed Consumer Advocate Retires." *Philadelphia Inquirer*, November 11, 2012.

McCarty, Nolan. 2014. "Complexity, Capacity, and Capture." In *Preventing Regulatory Capture*, edited by Daniel Carpenter and David A. Moss, 57–68. New York: Cambridge University Press.

Meckling, Jonas. 2011. *Carbon Coalitions*. Cambridge, MA: MIT Press.

Meckling, Jonas. 2015. "Oppose, Support, or Hedge? Distributional Effects, Regulatory Pressure, and Business Strategy in Environmental Politics." *Global Environmental Politics* 15 (2): 19–37.

Meckling, Jonas. 2019. "A New Path for US Climate Politics: Choosing Policies That Mobilize Business for Decarbonization." *The Annals of the American Academy of Political and Social Science* 685 (1): 82–95.

Mildenberger, Matto. 2020. *Carbon Captured: How Business and Labor Control Climate Politics*. Cambridge, MA: MIT Press.

Mildenberger, Matto. 2021. "The Development of Climate Institutions in the United States." *Environmental Politics* 30: 71–92.

Mildenberger, Matto, Jennifer R. Marlon, Peter D. Howe, and Anthony Leiserowitz. 2017. "The Spatial Distribution of Republican and Democratic Climate Opinions at State and Local Scales." *Climatic Change* 145: 539–548.

Miliband, Ralph. 1969. *The State in Capitalist Society*. New York: Basic Books.

Montañés, Cristina Crespo, Eric O'Shaughnessy, Dominic Bednar, Destenie Nock, Sanya Carley, Daniel M. Kammen, and David M. Konisky. 2023. "Enabling and Centering Equity and Justice in Clean Energy Transition Research." *Joule* 7: 431–449.

Moynihan, Donald P., and Joe Soss. 2014. "Policy Feedback and the Politics of Administration." *Public Administration Review* 74 (3): 320–332.

National Association of Attorneys General. 2024. "Powers and Duties." Accessed August 18, 2024. https://www.naag.org/issues/powers-and-duties/.

National Conference of State Legislatures. 2023. "State Renewable Portfolio Standards and Goals." Accessed August 2, 2023. https://www.ncsl.org/energy/state-renewable-portfolio-standards-and-goals.

Newmark, Adam J., and Christopher Witko. 2007. "Pollution, Politics, and Preferences for Environmental Spending in the States." *Review of Policy Research* 24 (4): 291–308.

Northwest Power and Conservation Council. 2023. "Dams: History and Purpose." https://www.nwcouncil.org/reports/columbia-river-history/damshistory/. Accessed December 3, 2023.

Oatley, Thomas. 2024. "The Dual Economy, Climate Change, and the Polarization of American Politics." *Socio-Economic Review* 22 (3): 1355–1380.

Oklahoma Attorney General. 2008. Comments in response to "Notice of Inquiry." Cause No. PUD 200800009.

Oklahoma Corporation Commission. 2008. "Notice of Inquiry." Cause No. PUD 200800009.

Oklahoma Corporation Commission. 2008a. "Final Order Approving Joint Stipulation and Settlement Agreement." PUD 200800086.

Oklahoma Electric Restructuring Advisory Committee. 2001. "Interim Report Relating to Transmission Issues." September–December 2001.

Oklahoma Gas and Electric. 2024. "Our Culture." Accessed June 1, 2024.

Oklahoma House Bill 2469. 2006. Died in Conference Committee, April 12, 2006. Would have Amended Section 2357.32A of Title 68. 2006 Regular Session.

Oklahoma House Bill 3028. 2010. Signed by Governor, May 27, 2010. 2010 Regular Session.

Oklahoma Industrial Energy Consumers. 2008. Comments in response to "Notice of Inquiry." Cause No. PUD 200800009.

Oklahoma Senate Bill 500. Signed by Governor, April 28, 1997. Oklahoma 1997–1998 Regular Legislative Session.

Oklahoma Senate Bill 220. 2000. Died in Conference Committee, May 26, 2000. Oklahoma 1999–2000 Regular Legislative Session.

Oklahoma Senate Bill 440. 2001. Signed by Governor, May 24, 2001. Creating New Section 190.20 of Title 17, Creating New Section 2357.32A of Title 68, Oklahoma Statutes. 2001 Regular Session.

Oregon House Bill 3039/Act 748. Signed by Governor, July 22, 2009. Chapter 748. Oregon Session Laws, 2009.

Oregon House Bill 3649/Act 71. 2010. Signed by Governor, March 18, 2010. Chapter 71. Oregon Session Laws, 2010.

Oregon House Bill 3674/Act 17. 2010. Signed by Governor, March 4, 2010. Chapter 17. Oregon Session Laws, 2010.

Oregon House Bill 3691/Act 79. 2010. Signed by Governor, March 18, 2010. Chapter 79.

Oregon Session Laws, 2010.

Oregon Senate Bill 1547/Act 28. 2016. Signed by Governor, March 8, 2016. Chapter 28. Oregon Session Laws, 2016.

Oregon House Bill 2021/Act 508. 2021. Signed by Governor, July 19, 2021. Chapter 508.

Oregon Session Laws, 2021.

Oregon Senate Bill 838/Act 301. 2007. Signed by Governor, June 13, 2007. Chapter 301. Oregon Session Laws, 2007.

Oregon Senate Bill 1149, C-Engrossed. 1999. Ordered by the House, June 29, 1999. 70th Oregon Legislative Assembly, 1999 Regular Session.

Page, Benjamin I., Jason Seawright, and Matthew J. Lacombe. 2019. *Billionaires and Stealth Politics*. Chicago: University of Chicago Press.

Parinandi, Srinivas C. 2023. *Following in Footsteps or Marching Alone? How Institutional Differences Influence Renewable Energy Policy*. Ann Arbor, MI: University of Michigan Press.

Pennsylvania Department of Environmental Protection. 2024. "Coal Mining in Pennsylvania." Accessed April 7, 2024. https://www.dep.pa.gov/Business/Land/Mining/Pages/PA-Mining-History.aspx#:~:text=Anthracite%20Coal,-Anthracite%20coal%20is&text=Anthracite%20mining%20is%20spread%20out,%2C%20Northumberland%2C%20and%20Luzerne%20counties?.

Pennsylvania House Bill 2300. 1996. Died in House Consumer Affairs Committee, January 2, 1996. Pennsylvania 1995–1996 Regular Legislative Session.

Pennsylvania House Bill 1509. 1996. Signed by Governor, December 3, 1996. Creating Chapter 28 of Title 66 of the Pennsylvania Consolidated Statutes. 1995–1996 Regular Session.

Pennsylvania House Bill 1203/Act 35. 2007. Signed by Governor, July 17, 2007. 2007–2008 Regular Session, Pennsylvania General Assembly.

Pennsylvania House Journal. 1996. 180th Session of the General Assembly. Monday, November 25, 1996. Pages 2553–2568.

Pennsylvania House Journal. 2004. 188th Session of the General Assembly. Saturday, November 20, 2004. Pages 2231–2320.

Pennsylvania Office of the Consumer Advocate. 2024. "About Us." Accessed April 15, 2024. https://www.oca.pa.gov/.

Pennsylvania Office of the Small Business Advocate. 2024. "Who We Are." Accessed April 15, 2024. https://www.osba.pa.gov/About/Pages/default.aspx.

Pennsylvania Public Utility Commission. 1995. "Initial Hearing." December 1, 1995. Harrisburg, PA. Investigation into Electric Power Competition, Docket I-00940032.

Pennsylvania Public Utility Commission. 1995a. "Uniform Cover and Calendar Sheet." Docketed October 10, 1995. Harrisburg, PA. Investigation into Electric Power Competition, Docket I-00940032.

Pennsylvania Public Utility Commission. 1995b. "Further Hearing." December 8, 1995. Harrisburg, PA. Investigation into Electric Power Competition, Docket I-00940032.

Pennsylvania Public Utility Commission. 1996. "Public Input Hearing." March 5, 1996.

Pittsburgh, PA. Investigation into Electric Power Competition, Docket I-00940032.

Pennsylvania Public Utility Commission. 1996a. "Public Input Hearing." January 30, 1996.

Philadelphia, PA. Investigation into Electric Power Competition, Docket I-00940032.

Pennsylvania Public Utility Commission. 1996b. "Further Hearing." January 16, 1996. Harrisburg, PA. Investigation into Electric Power Competition, Docket I-00940032.

Pennsylvania Public Utility Commission. 1996c. "Further Hearing." February 1, 1996. Harrisburg, PA. Investigation into Electric Power Competition, Docket I-00940032.

Pennsylvania Public Utility Commission. 1996d. "Further Hearing." March 19, 1996. Harrisburg, PA. Investigation into Electric Power Competition, Docket I-00940032.

Pennsylvania Public Utility Commission. 1997. "Application of PECO Energy Company for Approval of Its Restructuring Plan Under Section 2806 of the Public Utility Code." April 1, 1997. Docket R-00973953.

Pennsylvania Public Utility Commission. 1997a. "Opinion and Order on PECO Restructuring Plan." December 15, 1997. Docket R-00973953.

Pennsylvania Public Utility Commission. 1997b. "Pennsylvania Power and Light Company Restructuring Plan Filing." April 1, 1997. Docket R-00973954.

Pennsylvania Public Utility Commission. 1997c. "Metropolitan Edison Company Application for Approval of Restructuring Plan Under Section 2806 of the Public Utility Code." June 3, 1997. Docket R-00974008.

Pennsylvania Public Utility Commission. 1997d. "Pennsylvania Electric Company Application for Approval of Restructuring Plan Under Section 2806 of the Public Utility Code." June 3, 1997. Docket R-00974009.

Pennsylvania Public Utility Commission. 1997e. "Duquesne Light Company Application for Approval of Restructuring Plan Under Section 2806 of the Public Utility Code." August 1, 1997. Docket R-00974104.

Pennsylvania Public Utility Commission. 1997f. "UGI Utilities Application for Approval of Restructuring Plan Under Section 2806 of the Public Utility Code." August 22, 1997. Docket R-00973975.

Pennsylvania Public Utility Commission. 1997g. "Pennsylvania Power Company Application for Approval of Restructuring Plan Under Section 2806 of the Public Utility Code." October 7, 1997. Docket R-00974149.

Pennsylvania Public Utility Commission. 1998. "Opinion and Order on Pennsylvania Power and Light Company Restructuring Plan." June 19, 1998.

Pennsylvania Public Utility Commission. 1998a. "Opinion and Order on Metropolitan Edison Company Restructuring Plan." October 20, 1998. Docket R-00974008.

Pennsylvania Public Utility Commission. 1998b. "Opinion and Order on Pennsylvania Electric Company Restructuring Plan." October 20, 1998. Docket R-00974009.

Pennsylvania Public Utility Commission. 1998c. "Opinion and Order on Duquesne Light Company Restructuring Plan." May 30, 1998. Docket R-00974104.

Pennsylvania Public Utility Commission. 1998d. "Opinion and Order on UGI Utilities Restructuring Plan." June 29, 1998. Docket R-00973975.

Pennsylvania Public Utility Commission. 1998e. "Opinion and Order on Pennsylvania Power Company Restructuring Plan." July 27, 1998. Docket R-00974149.

Pennsylvania Public Utility Commission. 2005. "Implementation Order 1." March 23, 2005. Docket M-00051865.

Pennsylvania Public Utility Commission. 2005a. "Implementation Order 2." July 14, 2005. Docket M-00051865.

Pennsylvania Public Utility Commission. 2008. "Implementation Order." September 25, 2008. Docket L-00060180.

Pennsylvania Senate Bill 962. 2003. Died in Senate Consumer Protection and Professional Licensure Committee. 2003–2004 Regular Session, Pennsylvania General Assembly.

Pennsylvania Senate Bill 1030/Act 213. 2004. Signed by Governor, November 30, 2004. 2003–2004 Regular Session, Pennsylvania General Assembly.

Pennsylvania Senate Journal. 1996. 180th Session of the General Assembly. Monday, November 25, 1996. Pages 2674–2705.

Pennsylvania Senate Journal. 2004. 188th Session of the General Assembly. Wednesday, November 17, 2004. Pages 2298–2323.

Pennsylvania State House of Representatives Archives. 1995. "House Consumer Affairs Committee, 1995–1996." Box 4, Folder 1. Harrisburg, PA.

Pennsylvania State House of Representatives Archives. 2004. "Papers of L. Chris Ross, 1997–2016." Box 1, Folder 19. Harrisburg, PA.

Pennsylvania State House of Representatives Archives. 2007. "House Appropriations Committee Files, 2007–2008." Box 2, Folder 16. Harrisburg, PA.

Pennsylvania State House of Representatives Archives. 2009. "Papers of L. Chris Ross, 1997–2016." Box 2, Folder 1. Harrisburg, PA.

Pennsylvania State House of Representatives Archives. 2009a. "Papers of L. Chris Ross, 1997–2016." Box 2, Folder 2. Harrisburg, PA.

Pennsylvania State House of Representatives Archives. 2009b. "Papers of L. Chris Ross, 1997–2016." Box 2, Folder 3. Harrisburg, PA.

Pennsylvania State House of Representatives Archives. 2009c. "House Environmental Resources and Energy Committee Files, 2009–2010." Box 1, Folder 12. Harrisburg, PA.

Pennsylvania State House of Representatives Archives. 2009d. "House Environmental Resources and Energy Committee Files, 2009–2010." Box 1, Folder 14. Harrisburg, PA.

Pennsylvania State House of Representatives Archives. 2010. "Papers of L. Chris Ross, 1997–2016." Box 2, Folder 1. Harrisburg, PA.

Peskoe, Ari. 2023. "Replacing the Utility Transmission Syndicate's Control." *Energy Law Journal* 44 (3): 547–618.

Pierson, Paul. 1993. "When Effect Becomes Cause: Policy Feedback and Political Change." *World Politics* 45 (4): 595–628.

Pierson, Paul. 2000. "Increasing Returns, Path Dependence, and the Study of Politics." *American Political Science Review* 94 (2): 251–267.

Pierson, Paul, and Eric Schickler. 2024. *Partisan Nation: The Dangerous New Logic of American Politics in a Nationalized Era.* Chicago: University of Chicago Press.

Pomerantz, David. 2023. "Guess Who's Been Paying to Block Green Energy. You Have." *New York Times*, July 5, 2023.

Postelwait, Jeff. 2022. "Texas' Big Freeze: The 2021 Power Crisis and the Lessons Learned One Year Later." T&D World, January 25, 2022. https://www.tdworld.com/disaster-response/article/21213032/texas-big-freeze-the-2021-power-crisis-and-the-lessons-learned-one-year-later. Accessed July 4, 2023.

Poulantzas, Nicos. 1973. *Political Power and Social Classes*. London: New Left Books.

Power Technology. 2024. "Sooner Thermal Power Plant, US" Accessed June 1, 2024. https://www.power-technology.com/marketdata/sooner-thermal-power-plant-us/.

Prasad, Monica. 2023. "A Carbon Tax by Any Other Name: Public Benefit Funds in the American States." *PLOS Climate* 2 (2): e0000073.

Prechel, Harland. 2020. *Normalized Financial Wrongdoing.* Stanford, CA: Stanford University Press.

Prechel, Harland, and Theresa Morris. 2010. "The Effects of Organizational and Political Embeddedness on Financial Malfeasance in the Largest US Corporations: Dependence, Incentives and Opportunities." *American Sociological Review* 75 (3): 331–354.

Pritchard, Zachary D., and Sarah Mills. 2021. "Renewable Energy Requirements on the Ballot: An Analysis of County-Level Voting Results. *Energy Policy* 148: 111949.

Pub. L. 102–486. 102nd Congress, United States of America. https://www.congress.gov/bill/102nd-congress/house-bill/776/text. Accessed July 7, 2023.

Public Utility Commission of Texas. 1999. Substantive Rule §25.173. December 20, 1999.

Public Utility Commission of Texas. 2024. "Market Directories." Accessed January 8, 2024. https://www.puc.texas.gov/industry/electric/directories/default.aspx.

Public Service Company of Oklahoma. 2018. "PSO Facts." Accessed June 1, 2024. https://www.psoklahoma.com/lib/docs/cleanenergy/PSO2019FactSheet4-8-2019.pdf.

Rabe, Barry G. 2004. *Statehouse and Greenhouse: The Emerging Politics of American Climate Change Policy*. Washington, DC: Brookings Institution Press.

Rabe, Barry G., and Philip A. Mundo. 2007. "Business Influence in State-Level Environmental Policy." In *Business and Environmental Policy: Corporate Interests in the American Political System*, edited by Michael E. Kraft and Sheldon Kamieniecki. Cambridge, MA: MIT Press.

Rabe, Barry G. 2013. "Racing to the Top, the Bottom, or the Middle of the Pack? The Evolving State Government Role in Environmental Protection." In *Environmental Policy: New Directions for the 21st Century*, edited by N. Vig and M. Kraft, 30–53. Washington, DC: CQ Press.

Raymond, Leigh. 2010. "Beyond Additionality in Cap-and-Trade Offset Policy." *Issues in Governance Studies* 35 (July):1–9.

Raymond, Leigh. 2016. *Reclaiming the Atmospheric Commons: The Regional Greenhouse Gas Initiative and a New Model of Emissions Trading*. Cambridge, MA: MIT Press.

Rillinger, Georg. 2019. "Corporate Conspiracies and Complex Secrets: Structure and Perception of the Insull Scheme in 1930s Chicago." *American Journal of Sociology* 124 (4): 1043–1089.

Rilinger, Georg. 2023. "Who Captures Whom? Regulatory Misperceptions and the Timing of Cognitive Capture." *Regulation and Governance* 17 (1): 43–60.

Rilinger, Georg. 2024. *Failure by Design: The California Energy Crisis and the Limits of Market Planning*. Chicago: University of Chicago Press.

Roberts, David. 2024. "This New Organization Wants to Remake PUCs for the Energy Transition." *Volts Podcast*, September 11.

Rose, Kenneth, Robert E. Burns, Kenneth W. Costello, Robert J. Graniere, John C. Hoag, Mohammad Harunuzzaman, and Sridarshan Koundinya. 1997. "Summary of Key State Issues of FERC Orders 888 and 889." *The National Regulatory Research Institute*. Columbus, OH: Ohio State University.

Rosenthal, Alan. 2004. *Heavy Lifting: The Job of the American Legislature*. Washington, DC: CQ Press.

Rosenthal, Alan. 2009. *Engines of Democracy: Politics and Policymaking in State Legislatures*. Washington, DC: CQ Press.

Rountree, Valerie. 2019. "Nevada's Experience with the Renewable Portfolio Standard." *Energy Policy* 129: 279–291.

Schattschneider, E. E. 1935. *Politics, Pressure, and the Tariff*. New York: Prentice-Hall.

Schladen, Marty, and Morgan Trau. 2024. "Ex-First Energy Executives, Ohio Utility Regulator Charged by State in Bailout and Bribery Scandal." *Ohio Capital Journal*, February 12.

Schneider, Anne Larason, and Helen Ingram. 1997. *Policy Design for Democracy*. Lawrence, KS: University of Kansas Press.

Simmons, Erica S., and Nicholas Rush Smith. 2021. *Rethinking Comparison: Innovative Methods for Qualitative Political Inquiry*. New York: Cambridge University Press.

South Carolina Senate Bill 431/Act 16. 2007. Became Law Without Governor's Signature, May 3, 2007. South Carolina Laws, 2007.

South Carolina Senate Bill 1189/Act 236. 2014. Signed by Governor, June 2, 2014. Chapter 236. South Carolina Laws, 2014.

South Carolina House Bill 3659/Act 62. 2019. Signed by Governor, May 16, 2019. Chapter 62. South Carolina Laws, 2019.

South Carolina House Bill 4940/Act 187. 2020. Signed by Governor, September 29, 2020. Chapter 187. South Carolina Laws, 2020.

Spence, David B. 2024. *Climate of Contempt: How to Rescue the US Energy Transition from Voter Partisanship*. New York: Columbia University Press.

Stadler, Steve, J. Scott Greene, and Lauren Wood. 2017. "Oklahoma Wind Pushes Through Opposition." Presentation at Southwest Association of American Geographers Meeting.

Stigler, George J. 1971. "The Theory of Economic Regulation." *The Bell Journal of Economics and Management Science* 2: 3–31.

Stokes, Leah C. 2020. *Short Circuiting Policy: Interest Groups and the Battle Over Clean Energy and Climate Policy in the American States*. New York: Oxford University Press.

Stouffer, Rick. 2004. "Rendell Touts Clean Energy Legislation." *Pittsburgh Tribune-Review,* December 8, 2004.

Texas House Bill 1090. 2007. Signed by Governor, June 15, 2007. 80th Regular Session of the Texas Legislature.

Texas Legislature Online. 2024. "Search Legislation." Accessed January 8, 2024. https://capitol.texas.gov/.

Texas Senate Bill 373. 1995. Signed by Governor, June 16, 1995. Amending Public Utility Regulatory Act of 1995, as Enacted by SB 319. 74th Regular Session of the Texas Legislature.

Texas Senate Bill 7. 1999. Signed by Governor, June 18, 1999. Creating Chapters 39–41 of the Texas Utilities Code. 76th Regular Session of the Texas Legislature.

Texas Senate Bill 20. 2005. Signed by Governor, August 2, 2005. 79th Special Session #1 of the Texas Legislature.

Texas State Legislative Archives. 1999. Senate Floor Audio Recording, March 17, 1999.

Texas State Legislative Archives. 2005. "House Regulated Industries May 5, 2005 Meeting." Audio Recording.

Texas State Legislative Archives. 2009. "Senate Business and Commerce March 24, 2009 Meeting." Box 47, Folder 8. Austin, Texas.

The Editorial Board. 2024. "Failure of SDG&E Takeover an Indictment of City Hall." *San Diego Union-Tribune,* May 19.

The Oklahoman. 2000. "American Electric Power Acquires Owner of PSO." June 16, 2000. Accessed March 27, 2024. https://www.oklahoman.com/story/news/2000/06/16/american-electric-power-acquires-owner-of-pso/62193876007/.

Thombs, Ryan P., and Andrew K. Jorgenson. 2020. "The Political Economy of Renewable Portfolio Standards in the United States." *Energy Research and Social Science* 62: 101379.

Trabish, Herman K. 2019. "As California Utilities Falter, Legislators, Governor Back 'Tower of Finance' Solution to Wildfire Costs." *Utility Dive,* April 26.

Trachtman, Samuel. 2020. "What Drives Climate Policy Adoption in the US States?" *Energy Policy* 138: 111214.

Trachtman, Samuel, and Jonas Meckling. 2022. "The Climate Advocacy Gap." *Climatic Change* 172 (3–4): 24.

Travel Oklahoma. 2024. "Golden Driller." Accessed March 26, 2024. https://www.travelok.com/listings/view.profile/id.16498.

Truman, David B. 1955. *The Governmental Process.* New York: Knopf.

Tsoukalis, John H., Kathleen Spees, Johannes P. Pfeifenberger, Andrew Levitt, Andrew W. Thompson, Oleksandr Kuzura, Evan Bennett, Son Phan, Megan Diehl, Ellery Curtis, Sylvia Tang, and Ryan Nelson. 2023. "Assessment of Potential Market Reforms for South Carolina's Electricity Sector." *The Brattle Group*, April 27, 2023.

US Census Bureau. 2021. "Oregon: 2020 Census." Accessed March 26, 2024. https://www.census.gov/library/stories/state-by-state/oregon-population-change-between-census-decade.html.

US Census Bureau. 2021a. "South Carolina: 2020 Census." Accessed March 26, 2024. https://www.census.gov/library/stories/state-by-state/south-carolina-population-change-between-census-decade.html.

US Census Bureau. 2021b. "Texas: 2020 Census." Accessed March 26, 2024. https://www.census.gov/library/stories/state-by-state/texas-population-change-between-census-decade.html.

US Census Bureau. 2021c. "Oklahoma: 2020 Census." Accessed March 26, 2024. https://www.census.gov/library/stories/state-by-state/oklahoma-population-change-between-census-decade.html.

US Census Bureau. 2021d. "Pennsylvania: 2020 Census." Accessed March 26, 2024. https://www.census.gov/library/stories/state-by-state/pennsylvania-population-change-between-census-decade.html.

US Department of Energy. 2017. "Quadrennial Energy System Review: Transforming the Nation's Electricity System." Office of Policy, January 2017. Accessed June 27, 2023.

US Environmental Protection Agency. 2023. "Power Market Structure." https://www.epa.gov/green-power-markets/power-market-structure. Accessed July 4, 2023.

US Energy Information Administration. 1995. Form 861 Data.

US Energy Information Administration. 1996. "Electric Sales and Revenue, 1995." Office of Coal, Nuclear, Electric and Alternate Fuels, US Department of Energy, Washington, D.C.

US Energy Information Administration. 1997. "Electric Sales and Revenue, 1996." Office of Coal, Nuclear, Electric and Alternate Fuels, US Department of Energy, Washington, D.C.

US Energy Information Administration. 1998. "Electric Sales and Revenue, 1997." Office of Coal, Nuclear, Electric and Alternate Fuels, US Department of Energy, Washington, D.C.

US Energy Information Administration. 2001. Form 861 Data.

US Energy Information Administration. 2008. Form 861 Data.

US Energy Information Administration. 2021. Form 861 Data.

US Energy Information Administration. 2021a. "Annual Electric Power Industry Report- Form EIA-861." Accessed July 2, 2023.

US Energy Information Administration. 2022. "Electricity: Historical State Data." https://www.eia.gov/electricity/data/state/. Accessed February 6, 2023.

US Energy Information Administration. 2023. "State Energy Data System (SEDS): 1960–2021 (complete)." https://www.eia.gov/state/seds/seds-data-complete.php#state Selection. Accessed December 3, 2023.

Vachon, Stephan, and Frederic C. Menz. 2006. "The Role of Social, Political and Economic Interests in Promoting State Green Electricity Policies." *Environmental Science and Policy* 9 (7–8): 652–662.

Vaheesan, Sandeep. 2024. *Democracy in Power: A History of Electrification in the United States*. Chicago: University of Chicago Press.

Vaheesan, Sandeep. 2024a. "The Best Way to Fight Heat Waves and Outages Is to Green the Grid: The Fastest Way to Green the Grid, in Turn, Is for Government to Step In." *The New Republic*, July 17.

Vasseur, Michael. 2014. "Convergence and Divergence in Renewable Energy Policy Among US States from 1998 to 2011." *Social Forces* 92(4): 1637–1657.

Walmart. 2008. Comments in response to "Notice of Inquiry." Cause No. PUD 200800009.

Walton, Robert. 2019. "Split Apart PG&E? The Utility is Open to It but Warns Rates Would Likely Rise." *Utility Dive*, February 21.

Walton, Robert. 2019a. "California Gov. Newsom to Sign $21B Wildfire Liability Bill Today." *Utility Dive*, July 12.

Walton, Robert. 2019b. "PG&E, NextEra Face Off in Bankruptcy-related FERC Filings." *Utility Dive*, January 23.

Walton, Robert. 2019c. "California Regulators Approve Utility Wildfire Mitigation Plans; PUC President to Step Down." *Utility Dive*, May 31.

Weiss, Robert S. 1995. *Learning From Strangers: The Art and Method of Qualitative Interview Studies*. New York: Simon and Schuster.

Welton, Shelley, and Joel Eisen. 2019. "Clean Energy Justice: Charting an Emerging Agenda." *Harvard Environmental Law Review* 43: 307.

Wilson, James Q. 1980. *The Politics of Regulation*. New York: Basic Books.

Wiser, Ryan H., and Ole Langniss. 2001. "The Renewables Portfolio Standard in Texas: An Early Assessment." Lawrence Berkeley National Laboratory. Berkeley, CA.

Xcel Energy. 2024. "Our Power Supply." Accessed August 23, 2024. https://corporate.my.xcelenergy.com/s/energy/generating-power.

Yin, Haitao, and Nicholas Powers. 2010. "Do State Renewable Portfolio Standards Promote In-State Renewable Generation?" *Energy Policy* 38 (2): 1140–1149.

Zhou, Shan, Barry D. Soloman, and Marilyn A. Brown. 2024. "The Spillover Effect of Mandatory Renewable Portfolio Standards." *PNAS* 121 (25): e2313193121.

Index

Page numbers containing *n* refer to notes, first by page number then by note number.
Page numbers ending in *t* refer to tables.
Page numbers ending in *fig* refer to figures.

American and Comparative Environmental Policy

David M. Konisky, Megan Mullin, Kate O'Neill, series editors

Russell J. Dalton, Paula Garb, Nicholas P. Lovrich, John C. Pierce, and John M. Whiteley, *Critical Masses: Citizens, Nuclear Weapons Production, and Environmental Destruction in the United States and Russia*

Daniel A. Mazmanian and Michael E. Kraft, eds., *Toward Sustainable Communities: Transition and Transformations in Environmental Policy*

Elizabeth R. DeSombre, *Domestic Sources of International Environmental Policy: Industry, Environmentalists, and U.S. Power*

Kate O'Neill, *Waste Trading among Rich Nations: Building a New Theory of Environmental Regulation*

Joachim Blatter and Helen Ingram, eds., *Reflections on Water: New Approaches to Transboundary Conflicts and Cooperation*

Paul F. Steinberg, *Environmental Leadership in Developing Countries: Transnational Relations and Biodiversity Policy in Costa Rica and Bolivia*

Uday Desai, ed., *Environmental Politics and Policy in Industrialized Countries*

Kent Portney, *Taking Sustainable Cities Seriously: Economic Development, the Environment, and Quality of Life in American Cities*

Edward P. Weber, *Bringing Society Back In: Grassroots Ecosystem Management, Accountability, and Sustainable Communities*

Norman J. Vig and Michael G. Faure, eds., *Green Giants?: Environmental Policies of the United States and the European Union*

Robert F. Durant, Daniel J. Fiorino, and Rosemary O'Leary, eds., *Environmental Governance Reconsidered: Challenges, Choices, and Opportunities*

Paul A. Sabatier, Will Focht, Mark Lubell, Zev Trachtenberg, Arnold Vedlitz, and Marty Matlock, eds., *Swimming Upstream: Collaborative Approaches to Watershed Management*

Sally K. Fairfax, Lauren Gwin, Mary Ann King, Leigh S. Raymond, and Laura A. Watt, *Buying Nature: The Limits of Land Acquisition as a Conservation Strategy, 1780–2004*

Steven Cohen, Sheldon Kamieniecki, and Matthew A. Cahn, *Strategic Planning in Environmental Regulation: A Policy Approach That Works*

Michael E. Kraft and Sheldon Kamieniecki, eds., *Business and Environmental Policy: Corporate Interests in the American Political System*

Joseph F. C. DiMento and Pamela Doughman, eds., *Climate Change: What It Means for Us, Our Children, and Our Grandchildren*

Christopher McGrory Klyza and David J. Sousa, *American Environmental Policy, 1990–2006: Beyond Gridlock*

John M. Whiteley, Helen Ingram, and Richard W. Perry, eds., *Water, Place, and Equity*

Judith A. Layzer, *Natural Experiments: Ecosystem-Based Management and the Environment*

Daniel A. Mazmanian and Michael E. Kraft, eds., *Toward Sustainable Communities: Transition and Transformations in Environmental Policy, second edition*

Henrik Selin and Stacy D. VanDeveer, eds., *Changing Climates in North American Politics: Institutions, Policymaking, and Multilevel Governance*

Megan Mullin, *Governing the Tap: Special District Governance and the New Local Politics of Water*

David M. Driesen, ed., *Economic Thought and U.S. Climate Change Policy*

Kathryn Harrison and Lisa McIntosh Sundstrom, eds., *Global Commons, Domestic Decisions: The Comparative Politics of Climate Change*

William Ascher, Toddi Steelman, and Robert Healy, *Knowledge and Environmental Policy: Re-Imagining the Boundaries of Science and Politics*

Michael E. Kraft, Mark Stephan, and Troy D. Abel, *Coming Clean: Information Disclosure and Environmental Performance*

Paul F. Steinberg and Stacy D. VanDeveer, eds., *Comparative Environmental Politics: Theory, Practice, and Prospects*

Judith A. Layzer, *Open for Business: Conservatives' Opposition to Environmental Regulation*

Kent E. Portney, *Taking Sustainable Cities Seriously: Economic Development, the Environment, and Quality of Life in American Cities, second edition*

Raul Lejano, Mrill Ingram, and Helen Ingram, *The Power of Narrative in Environmental Networks*

Christopher McGrory Klyza and David J. Sousa, *American Environmental Policy: Beyond Gridlock, updated and expanded edition*

Andreas Duit, ed., *State and Environment: The Comparative Study of Environmental Governance*

Joseph F. C. DiMento and Pamela Doughman, eds., *Climate Change: What It Means for Us, Our Children, and Our Grandchildren, second edition*

David M. Konisky, ed., *Failed Promises: Evaluating the Federal Government's Response to Environmental Justice*

Leigh Raymond, *Reclaiming the Atmospheric Commons: The Regional Greenhouse Gas Initiative and a New Model of Emissions Trading*

Robert F. Durant, Daniel J. Fiorino, and Rosemary O'Leary, eds., *Environmental Governance Reconsidered: Challenges, Choices, and Opportunities, second edition*

James Meadowcroft and Daniel J. Fiorino, eds., *Conceptual Innovation in Environmental Policy*

Barry G. Rabe, *Can We Price Carbon?*

Kelly Sims Gallagher and Xiaowei Xuan, *Titans of the Climate: Explaining Policy Process in the United States and China*

Matto Mildenberger, *Carbon Captured: How Business and Labor Control Climate Politics*

Mary Alice Haddad, *Effective Advocacy: Lessons from East Asia's Environmentalists*

George Hoberg, *The Resistance Dilemma: Place-Based Movements and the Climate Crisis*

Roger Karapin and David Vogel, *When Federal Climate Policy Works: Benefits, Business, and Politics*

Jin Sato, *Sustainability Inverted: How Environmental Policies Control People*

Joshua Basseches, *Owning the Green Grid: The Political Economy of Renewable Energy Policy Design*

Christopher McGrory Klyza, *Building the American Green State: The States, the Federal Government, and Intercurrence, 1800–2000*

Geoffrey Henderson, *Getting to Work on Climate Policy: The Strategy of Labor-Environmental Coalitions*

Publisher contact:
The MIT Press
Massachusetts Institute of Technology
77 Massachusetts Avenue, Cambridge, MA 02139
mitpress.mit.edu

EU Authorised Representative:
Easy Access System Europe, Mustamäe tee 50,
10621 Tallinn, Estonia
gpsr.requests@easproject.com

Printed by Integrated Books International,
United States of America